133
Advances in Polymer Science

Springer-Verlag Berlin Heidelberg GmbH

Metal Complex Catalysts
Supercritical Fluid Polymerization
Supramolecular Architecture

With contributions by
D. A. Canelas, J. M. DeSimone, A. Harada,
E. Ihara, K. Mashima, A. Nakamura,
Y. Nakayama, H. Yasuda

Springer

This series presents critical reviews of the present and future trends in polymer and biopolymer science including chemistry, physical chemistry, physics and materials science. It is addressed to all scientists at universities and in industry who wish to keep abreast of advances in the topics covered.

As a rule, contributions are specially commissioned. The editors and publishers will, however, always be pleased to receive suggestions and supplementary information. Papers are accepted for „Advances in Polymer Science" in English.

In references Advances in Polymer Science is abbreviated Adv. Polym. Sci. and is cited as a journal.

Springer WWW home page: http://www.springer.de

ISSN 0065-3195
ISBN 978-3-662-14793-1   ISBN 978-3-540-68442-8 (eBook)
DOI 10.1007/978-3-540-68442-8
Library of Congress Catalog Card Number 61642

© Springer-Verlag Berlin Heidelberg 1997

Originally published by Springer-Verlag Berlin Heidelberg New York in 1997
Softcover reprint of the hardcover 1st edition 1997

Typesetting: Macmillan India Ltd., Bangalore-25
Cover: E. Kirchner, Heidelberg
SPIN: 10548301   02/3020 - 5 4 3 2 1 0 - Printed on acid-free paper

Editors

Table of Contents

Recent Trends in the Polymerization of α-Olefins Catalyzed by Organometallic Complexes of Early Transition Metals

Kazushi Mashima[1], Yuushou Nakayama[2] and Akira Nakamura[2,*]

[1]Department of Chemistry, Graduate School of Engineering Science, Osaka University, Toyonaka, Osaka 560, Japan
[2]Department of Macromolecular Science, Graduate School of Science, Osaka University, Toyonaka, Osaka 560, Japan. *E-mail: anaka@chem.sci.osaka-u.ac.jp

This review article describes recent progress in the field of homogeneous organometallic catalysts for olefin polymerization and focuses on the metal-carbon bonding character of the transition metal complexes used as catalysts. Most catalysts of this kind are based on metallocene derivatives of Group 4 metals, their catalytic behavior (such as activity and stereospecificity) and the molecular weights of the resulting polymers are surveyed on the basis of the molecular structure of the catalyst precursors. Advanced mechanistic studies on the catalyst systems are also summarized. Some examples of the related polymerization of functionalized olefins are also presented. Not only the Group 4 metal catalysts but also the polymerization catalysts of many other early transition metals and late transition metals are reviewed including our recent study on the mono(cyclopenta-dienyl)mono(diene) complexes of Group 5 metals.

1 Introduction

The polymerization of olefins and di-olefins is one of the most important targets in polymer science. This review article describes recent progress in this field and deals with organo-transition metal complexes as polymerization catalysts. Recent developments in organometallic chemistry have prompted us to find a precise description of the mechanism of propagation, chain transfer, and termination steps in the homogeneously metal-assisted polymerization of olefins and diolefins. Thus, this development provides an idea for designing any catalyst systems that are of interest in industry.

Recently, the "agostic" interaction of alkyl group(s) on transition metals has emerged as highly basic and new concept and is found to be important in understanding the mechanism of the metal-catalyzed homogeneous oligomerization and polymerization of α-olefins. Early transition metal alkyl complexes generally have partially ionic M–C bonds and show *α-agostic hydrogen interaction* that somewhat stabilizes the catalytically active species by providing electrons at a vacant site on the metal. This is in sharp contrast to the fact that late transition metal alkyl complexes show mainly *β-agostic hydrogen interaction* that causes the hydrogen transfer easily through β-hydrogen elimination and reductive elimination, and that gives rise to the oligomerization of olefins. Organometallic complexes of the early and late transition metals have been used as catalysts for olefin oligomerization and polymerization. The mechanism involved in these catalyst systems depends very much upon the kind of metal centers as well as their co-ligands, and thus the different mechanisms which can be distinguished by detailed investigations should be assumed for early and late transition metal catalysts.

In this contribution, we review the mechanism of polymerization and oligomerization involving early transition metals, taking as our basis recent results in advanced organometallic chemistry. First of all, some recent examples of the previous reviews concerning the Ziegler-Natta polymerization are cited [1–10]. Then, relevant new reports are surveyed in a systematic fashion.

2 General Features of Organometallic Complexes of Early Transition Metals

Recently, a deeper understanding of the precise nature of metal–carbon bonding was achieved, enabling specific polymerization catalyst systems to be designed on a practical level. The metal–carbon bond of early transition metals is partially ionic, while that of late transition metal is generally covalent. The degree of ionicity is delicately dependent on the identity of metal, formal oxidation states and auxiliary ligands.

Table 1. Crystal and spectral data for some alkene complexes of early transition metals

Complex	M-C (Å)	C-C (Å)	C-M-C (deg)	$\delta(^{13}C)$ (ppm)	J_{C-H} (Hz)	Ref.
$CH_2=CH_2$	–	1.337(2)	–	123.3	156.4	[22–24]
Mononuclear complexes						
$Cp_2^*Ti(C_2H_4)$ (6)	2.160(4)	1.438(5)	38.9(1)	105.1	143.6	[25]
$Cp_2^*Zr(PhCH=CHPh)(PMe_3)$ (7)	2.36(2) 2.43(2)	1.38(2)	33.4(5)	–	–	[26]
$Cp_2Zr(CH_2=CH_2)(PMe_3)$ (8)	2.344(8) 2.373(8)	1.486(8)	36.7(3)	–	–	[27]
$Cp_2Zr(1\text{-butene})(PMe_3)$ (9)	2.35 2.37	1.44	35.5	–	–	[28]
$Cp_2Hf(CH_2=CMe_2)(PMe_3)$ (10)	2.316(8) 2.368(9)	1.46(1)	36.2(3)	–	–	[29]
$[Li(tmeda)]_2[Hf(CH_2=CH_2)_2Et_4]$ (11)	2.26(4) 2.31(4)	1.49(6)	38.0	–	–	[30]
$Cp_2Nb(C_2H_5)(C_2H_4)$ (12)	2.277(9) 2.320(9)	1.406(13)	35.6(3)	29.35 27.63	153 154.5	[31]
endo-$Cp_2^*Nb(H)(CH_2=CHPh)$ (13)	2.289(4) 2.309(4)	1.431(6)	36.3(1)	–	–	[32]
$Cp_2Nb(C_2H_4)(SiMe_3)$ (14)	2.314(3) 2.317(3)	1.440(5)	36.2(1)	13.5 14.0	–	[33]
$CpNb(=NC_6H_3Pr^i_2\text{-}2,6)(PMe_3)(CH_2=CHMe)$ (15)	2.39(3) 2.28(3)	1.58(4)	39.4	–	–	[34]

Compound	M–C	C–C				Ref.
Cp*Ta(CHCMe₃)(PMe₃)(C₂H₄) (16)	2.228(3), 2.285(3)	1.477(4)	38.1(9)	–	–	[35]
TaCl₂{C₆H₃(CH₂NMe₂)₂-2,6}{CH₂=CHBuᵗ} (17)	2.200(7), 2.211(7)	1.436(10)	38.0(3)	95.2, 87.6	145.5, 139.2	[36]
Bridging alkene complexes						
(Cp₂ZrClAlEt₃)₂(CH₂CH₂) (18)	2.36, 2.49	1.55	37.2	–	–	[16]
(Cp₂ZrMe)₂(CH₂CH₂) (19)	2.327(6), 2.528(4)	1.473(1)	35.1(2)	–	–	[17]
{ZrCl₃(PEt₃)₂}₂(CH₂CH₂) (20)	2.42(2), 2.44(2)	1.69(3)	40.6(6)	–	–	[18]
{ZrBr₃(PEt₃)₂}₂(CH₂CH₂) (21)	2.41(2), 2.40(2)	1.56(3)	37.9(7)	–	–	[18]
{HfCl₃(PEt₃)₂}₂(CH₂CH₂) (22)	2.386(10), 2.364(9)	1.476(14)	36.2(3)	–	–	[18]
{HfBr₃(PEt₃)₂}₂(CH₂CH₂) (23)	2.36(2), 2.374(15)	1.51(2)	37.1(5)	–	–	[18]
Cp*₂Yb(μ-C₂H₄)Pt(PPh₃)₂ (4)	2.770(3), 2.793(3)	1.436(5)	29.9(1)	–	–	[19]
Cp*₂Sm(μ-η²:η²-CH₂CHPh)SmCp*₂ (5)	2.537(15), 2.647(15), 2.674(15), 2.732(15)	1.468(22)	32.8(5), 31.5(5)	–	–	[20]

As a typical case, olefin-metal complexation is described first. Alkene complexes of d^0 transition metals or ions have no d-electron available for the π-back donation, and thus their metal–alkene bonding is too weak for them to be isolated and characterized. One exception is $Cp_2^*YCH_2CH_2C(CH_3)_2CH=CH_2$ (1), in which an intramolecular bonding interaction between a terminal olefinic moiety and a metal center is observed. However, this complex is thermally unstable above $-50\,°C$ [11]. The MO calculation proves the presence of the weak metal-alkene bonding during the propagation step of the olefin polymerization [12, 13].

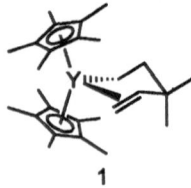

1

As shown in Table 1, a remarkable variety of alkene complexes bearing metal centers in a low oxidation state have been isolated and their structures have been determined by X-ray analysis. All the C–C bond distances in olefins coordinated to early transition metals at low oxidation states are more or less elongated compared to free ethylene. These structural data, together with those from NMR studies [14], indicate a major contribution of the metallacyclopropane structure (2), a fact which is also supported by calculation studies [15]. In the case of ethylene bridging two metal centers such as $[\{Cp_2ZrX\}_2(\mu\text{-}\eta\text{-}C_2H_4)]$ (3), the M–C bond could be characterized as a σ-bond and there is a little contribution from the μ-ethylene canonical structure [16–18].

$$L_nM\!\!<\!\!\rfloor \qquad\qquad L_nM\diagdown\diagup ML_n \;\longleftarrow\; L_nM\text{---}\|\text{---}ML_n$$

2   **3**

Organometallic complexes of the lanthanide series have strong Lewis acidic sites, in σ and π fashion. In particular, η^2-interaction with C=C bonds has special importance in the mechanism of polymerization. However, only a few examples are known for isolated compounds such as $Cp_2^*Yb(\mu\text{-}C_2H_4)Pt(PPh_3)_2$ (4) [19], and $Cp_2^*Sm(\mu\text{-}\eta^2\text{:}\eta^4\text{-}CH_2CHPh)SmCp_2^*$ (5) [20]. An acetylene adduct of ytterbium, $Cp_2^*Yb(MeC\equiv CMe)$, has been characterized crystallographically and shown to have weak interaction, like a Lewis acid–base interaction, and no π-back bonding [21].

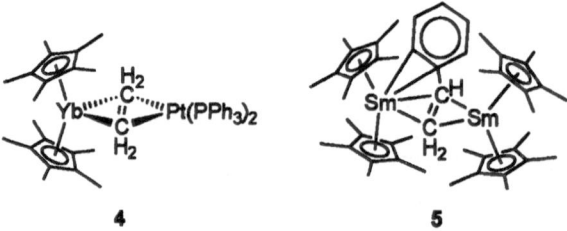

4   5

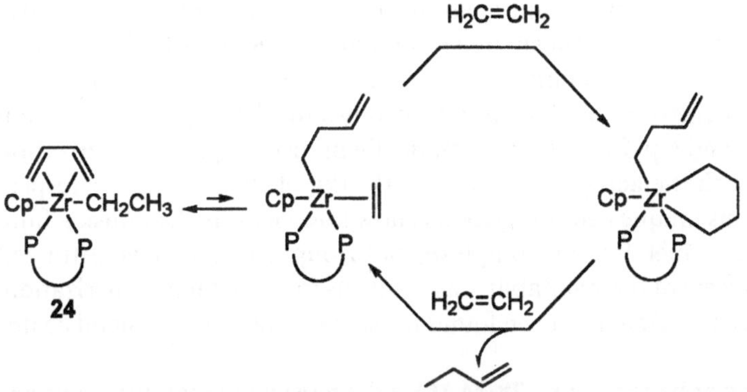

Scheme 1.

The π-back donation stabilizes the alkene-metal π-bonding and therefore this is the reason why alkene complexes of the low-valent early transition metals so far isolated did not catalyze any polymerization. Some of them catalyze the oligomerization of olefins via metallocyclic mechanism [25, 30, 37–39]. For example, a zirconium-alkyl complex, $CpZr^{II}(CH_2CH_3)(\eta^4$-butadiene)(dmpe) (dmpe = 1,2-bis(dimethylphosphino)ethane) (**24**), catalyzed the selective dimerization of ethylene to 1-butene (Scheme I) [37, 38].

Dimethyltitanium complex **25**, bearing an ethylene and methyl ligands, catalyzed the dimerization of ethylene via a metallacyclopentane intermediate **26** (Eq. 1) [30]. During the dimerization, no insertion of ethylene into the Ti–Me bond was observed due to the perpendicular orientation between methyl and ethylene ligands. This inertness could be attributed to the low oxidation state of **25**, i.e. Ti(II).

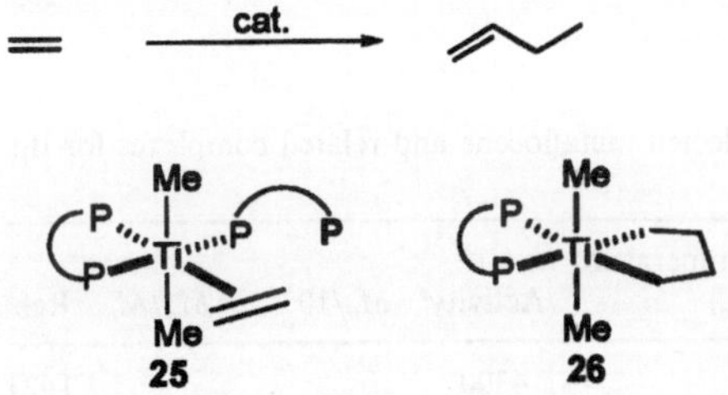

(1)

3 Olefin Polymerization by Organometallic Complexes

3.1 Cationic Metallocene Complexes as Active Species for a Metallocene/MAO System

A catalyst system consisting of Cp_2TiCl_2 (Cp = η^5-cyclopentadienyl) (**27**) and alkylaluminum, such as $AlClEt_2$, showed only a low activity for α-olefin polymerization [40]. The important breakthrough in this homogeneous catalyst system was reached serendipitously by H. Sinn and W. Kaminsky [41] during a study of the effect of methylaluminoxane (abbreviated as MAO) on polymerization. Addition of an excess of MAO as a co-catalyst to a homogeneous solution of Cp_2TiCl_2 (27) and Cp_2ZrCl_2 (28) dramatically induced rapid polymerization of ethylene to reach the activities of 10^4 and 10^5 (kg of PE/mol.h.atm), respectively [42,43]. This unique catalyst system is now known as the Kaminsky–Sinn catalyst. Since then metallocene complexes of Group 4 metals have attracted much interest in view of their suitability as catalysts for α-olefin polymerization; and the zirconocene system has been found to have the most active metal center [42–44].

For propylene polymerization, 28/MAO is the most active catalyst component in reactions that yield atactic polypropylene. The activity of a catalyst system of hafnocene/MAO is generally lower than that of the zirconocene catalyst systems, but the former has the advantage that it yields polymers of higher molecular weight [45,46]. Some typical activity values of selected metallocene systems catalyzing the homo- and co-polymerization of α-olefins are shown in Tables 2 and 3, even though the activity, stereospecificity and molecular weight of the resulting polymers significantly depend on the polymerization conditions, i.e., monomer and catalyst concentration as well as the amount and kind of the co-catalyst used. In fact, the polymerization of propylene using rac-$(C_2H_4)(IndH_4)_2ZrCl_2$ (29)/MAO and rac-$(C_2H_4)(Ind)_2ZrCl_2$ (30)/MAO as catalysts often leads to results that are at variance with literature data; this is because stereospecificity is closely related to monomer concentration [47,48].

The electronic and steric effects of substituents on cyclopentadienyl ligands that affect the activity of the catalyst have been extensively and intensively

Table 2. Examples of the catalytic activity of selected metallocene and related complexes for the polymerization of ethylene

Complex	Temperature (°C)	Activity[a]	$M_n/10^3$	M_w/M_n	Ref.
Cp_2TiCl_2 (27)/MAO	20	4300			[42]
Cp_2TiMe_2 (32)/MAO	20	9600	$530(M_\eta)$		[43]
$S(2,2'\text{-}C_6H_2Bu^t\text{-}2\text{-}Me\text{-}4\text{-}O)_2TiCl_2$ (31)/MAO	20	39300	1800	2	[52]
Cp_2ZrCl_2 (28)/MAO	70	91000			[42]
Cp_2ZrMe_2 (33)/MAO	50	9800	331	1.78	[44]
$(Ind)_2ZrMe_2$ (34)/MAO	50	66000	395	2.20	[44]
$Me_2Si(Ind)_2ZrCl_2$ (35)/MAO	65	3900	201	2.8	[44]
$[Cp_2ZrMe][(C_2B_9H_{11})_2Co]$ (36)	60	400	3.5–9.4	3.3–11.6	[54]
$[Cp^*(C_2B_9H_{11})M(Me)]_x(M=Zr$ (37), Hf (38))	20	72			[55]
Cp_2HfMe_2 (39)/MAO	50	1700	361		[44]
Cp_2^*LaH (40)	25	182000	680	2.03	[56]
Cp_2^*LuMe (41)	50–80	6900	–	–	[57]
$[PhC(NSiMe_3)_2]_2ZrCl_2$ (42)/MAO	25	570	$162(M_\eta)$		[58]

[a]Activity: kg(mol of catalyst)$^{-1}$h^{-1}atm^{-1}.

Table 3. Catalytic activity of selected metallocene complexes activated with MAO for the polymerization of propylene

Complex	Temperature (°C)	Activity[a]	$M_n/10^3$	M_w/M_n	Tacticity	Ref.
Cp_2ZrMe_2 (33)	60	2730			atactic	[59]
rac-$(C_2H_4)(IndH_4)_2ZrCl_2$ (29)	65	7700	12	2.2	(mm 0.95)	[59]
rac-$Me_2Si(Ind)_2ZrCl_2$ (35)	30	110	201	2.0	mm 0.962	[53]
$Me_2Si(C_5HMe_3$-2,3,5$)(C_5HMe_3$-2',4',5'$)ZrCl_2$ (43)	50	5300	67	2.5	mmmm 0.977	[60]
$Me_2Si(C_5H_2Bu^t$-3-Me-5$)(C_5H_3Me$-2'-Bu^t-4'$)ZrCl_2$ (44)	40	505	3.7		mmmm 0.94	[61]
$(C_2H_4)(C_5H_2Bu^t$-3$)_2ZrCl_2$ (45) (42% rac 58% meso)		160	6.9[b]	2.5[b]	mmmm 0.976[b]	[62]
rac-$Me_2Si(Benz[e]Indenyl)_2ZrCl_2$ (46)	50	41100	24	1.64	mmmm 0.90	[63]
rac-$Me_2Si($2-Me-$Benz[e]Indenyl)_2ZrCl_2$ (47)	40	29000	114	1.71	mmmm 0.93	[63]
$Me_2C(Cp)(Flu)ZrCl_2$ (48)	25	20000	70	1.9	rrrr 0.86	[64]
Cp_2TiMe_2 (32)	20	36	88 (M_η)		atactic	[43]
Cp_2TiPh_2 (49)	−60	54	55	1.7	mmmm 0.53	[65]
$(C_2H_4)(Ind)_2TiCl_2$ (50) (56% rac 44% meso)	−60	18	97	1.6	mm 0.54	[65]
rac-$MeCH(Ind)(C_5Me_4)TiCl_2$ (51)	50	250	67	1.9	mmmm 0.40	[66]
rac-$MeCH(Ind)(C_5Me_4)TiMe_2$ (52)	25	8910		2.2	mmmm 0.38	[67]
$S(2,2'$-$C_6H_4Bu^t$-2-Me-4-O-1$)_2TiCl_2$ (31)	20		>4000		atactic	[52]
rac-$C_2H_4(Ind)_2HfCl_2$ (53)	50	26800	>724	2.2		[45]
rac-$(C_2H_4)(IndH_4)_2HfCl_2$ (54)	80	34800	42	2.4		[45]

[a] Activity: kg(mol of catalyst)$^{-1}$ h^{-1} atm^{-1}.
[b] Data of pentane insoluble fraction.

investigated. The presence of electron-withdrawing substituents on the auxilary ligands decreases the activity of the catalysts for ethylene and propylene polymerization and lowers the molecular weight of the resulting polymer [49, 50].

Although cyclopentadienyl ligands have been well investigated, the question of alternative ligands for olefin polymerization is relatively unexplored. Schaverien and coworkers have reported olefin polymerizations using a variety of Group 4 metal complexes bearing chelating aryloxide ligands [51]. But most of them are less active for polymerization of ethylene than metallocene catalysts, except for $S(2,2'\text{-}C_6H_2Bu^t\text{-}2\text{-}Me\text{-}4\text{-}O)_2TiCl_2$ (**31**) which had been reported by Miyatake et al. [52].

Although the role of MAO remains still speculative, organometallic approaches have led to progress in distinguishing catalytically active species [68]. Cationic alkyl metallocene complexes are now considered the catalytically active species in metallocene/MAO systems. Spectroscopic observation has confirmed the presence of cationic catalytic centers. X-ray photoelectron spectroscopy (XPS) on the binding energy of $Zr(3d_{5/2})$ has suggested the presence of cationic species, and cationic hydride species such as $ZrHCp_2^+$ that are generated by β-hydride elimination of the propagating chain end [69].

Marks et al. observed cationic species in solution by NMR spectroscopy. The transfer of the methyl group from Zr to Al and the formation of the cationic species $[ZrMeCp_2]^+$ were directly detected by measuring the CPMAS NMR spectra of powdery samples of Cp_2ZrMe_2/MAO, that were obtained after the evaporation of the solvent indicated [70–74]. Similarly, $Cp_2'AnR_2$ and ZrR_2Cp_2 in Lewis-acid on surfaces such as dehydroxylated γ-alumina afforded catalytically active cationic species as heterogeneous surface catalysts, which were also detected by CPMAS NMR spectroscopy. In contrast, organoactinide species supported by dehydroxylated silica afforded catalytically inactive $L_nAn\text{-}OSi$ species. These results indicate that the coordinative unsaturation and electrophilic character at the metal center are essential for the active catalyst.

The active species of the metallocene/MAO catalyst system have now been established as being three-coordinated cationic alkyl complexes $[Cp_2MR]^+$ (14-electron species). A number of cationic alkyl metallocene complexes have been synthesized with various anionic components. Some structurally characterized complexes are presented in Table 4 [75, 76]. These cationic Group 4 complexes are coordinatively unsaturated and often stabilized by weak interactions, such as agostic interactions, as well as by cation-anion interactions. Under polymerization conditions such weak interactions smoothly provide the metal sites for monomers.

In 1985, Eisch et al. isolated a cationic alkenyltitanium complex (**55**) by the insertion of an alkyne into the cationic Ti–C bond generated from titanocene dichloride and methylaluminum dichloride (Eq. 2) [77]. Similarly, a mixture of $Cp_2TiCl(CH_2SiMe_3)$ and $AlCl_3$ afforded the solvent-separated ion pairs,

Table 4. Bond lengths for cationic metallocene complexes

Complex	M-C (Å)	M$\cdots$C^a (Å)	Ref.
[Cp₂Ti{C(SiMe₃)=CPhMe}][AlCl₄] (55)	2.13	—	[77]
[Cp₂*TiMe(thf)][BPh₄] (56)	1.988(10)	—	[78]
[(C₅H₃Me₂-1,2)₂ZrMe][MeB(C₆F₅)₃] (57)	2.252(4)	2.549(3) (cation-anion)^b	[79]
[(C₅H₃(SiMe₃)₂-1,3)₂ZrMe][MeB(C₆F₅)₃] (58)	2.260(4)	2.667(5) (cation-anion)^b	[80]
[Cp₂ZrMe(thf)][BPh₄] (59)	2.256(10)	—	[81]
[Cp*₂Zr(CH₂SiMe₃)(thf)][BPh₄] (60)	2.238(6)	—	[82]
[Cp₂*ZrMe(tht)][BPh₄] (61)	2.242(8)	—	[83]
[Me₂C(Cp)(Flu)ZrMe(PMe₃)][B(C₆F₅)₄] (62)	2.23(2)	2.648(6) (C$_{ipso}$)^c	[84]
[Cp₂Zr(η²-CH₂Ph)(NCMe)][BPh₄] (63)	2.344(8)	2.629(9) (β-H agostic)^d	[85]
[(C₅H₄Me)₂Zr(CH₂CH₃)(PMe₃)][BPh₄] (64)	2.290(9)	2.57(2) (C$_\beta$-Si)^e	[86]
[(C₅H₄Me)₂Zr(CH₂CH₂SiMe₃)(thf)][BPh₄] (65)	2.26(2)	2.39 (C$_\beta$-Al)^g	[87]
[Cp₂Zr{CH₂CH(AlEt₂)₂}][C₅H₅] (66)	2.27		[16]
[rac-(C₂H₄)(Ind)₂Zr{η²-C,C-CH(SiMe₂Cl)(SiMe₃)}][Al₂Cl₆.₅Me₀.₅] (67)	2.300(5)	(2.573(1) (Zr-Cl))^h (3.132(1) (Zr-Si))^h	[88]
[(C₅H₄Bu^t)₂Zr{η⁵-CH₂C(Me)=C(Me)C(Me)=CH(Me)}][B(C₆H₄F-4)₄] (68)	2.315(7)	2.705(6) (Zr-C$_\beta$)^i 2.712(6) (Zr-C$_\gamma$)^i 2.602(7) (Zr-C$_\delta$)^i 2.759(7) (Zr-C$_\varepsilon$)^i	[89]
[Cp*₂HfMe(tht)][BPh₄] (69)	2.233(9)	(αH-agostic)	[83]
[Cp₂*Hf(CH₂CHMe₂)(thf)][BPh₄] (70)	2.25(1)	(2.648(8) (Hf-H$_\alpha$))^j (3.04(8) (Hf-H$_\alpha$))^j	[90]

^{a} Unusual metal-carbon contacts unless noted in parenthesis.
^{b} Interactions between the cationic metal center and the CH₃ group of the anion.
^{c} Zr-C$_{ipso}$ bond of Zr(η²-benzyl) group.
^{d} Zr-C$_\beta$ bond by β-hydrogen agostic interaction.
^{e} Zr-C$_\beta$ interaction by γ-silicon effect.
^{g} Zr-C$_\beta$ interaction presumably by γ-aluminium effect.
^{h} Zr-Cl, and Zr-Si$_\beta$ contacts, respectively.
^{i} Distances between the zirconium atom and the olefinic carbon atoms.
^{j} Hf-H$_\alpha$ distances connected by α-H agostic interaction.

Ti(CH$_2$SiMe$_3$)Cp$_2^+$ and AlCl$_4^-$, which were detected by NMR spectroscopy [91,92].

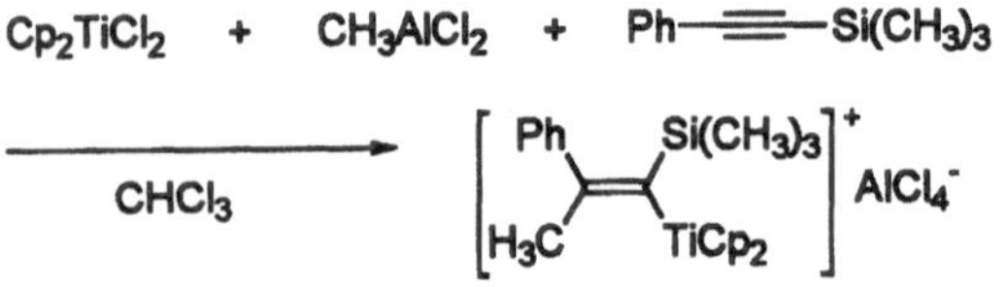

Cp$_2$TiCl$_2$ +   CH$_3$AlCl$_2$ +   Ph━━━Si(CH$_3$)$_3$

$$\xrightarrow{\text{CHCl}_3} \begin{bmatrix} \text{Ph} \quad\quad \text{Si(CH}_3)_3 \\ \diagup\!\!\!\!=\!\!\!\!\diagdown \\ \text{H}_3\text{C} \quad\quad \text{TiCp}_2 \end{bmatrix}^+ \text{AlCl}_4^-$$

(2)

55

Eisch's work promoted investigation into the preparation of cationic metallocene complexes of Group 4 metals. Several preparative routes to cationic group 4 metallocene complexes are illustrated in Scheme II. Catalytic activities of some selected cationic metallocene complexes for the polymerization of α-olefins are summarized in Tables 5 and 6. The catalyst systems based on these cationic complexes are just as active as MAO-activated metallocene catalysts for the polymerization of α-olefins.

Protonolysis of Cp$_2$TiMe$_2$ with HBF$_4$OEt$_2$ and NH$_4$X (X = PF$_6$, ClO$_4$) yielded the insoluble complexes $\{[$Cp$_2$TiMe]BF$_4\}_n$ (**86**) and [Cp$_2$TiMe(NH$_3$)]X (**87**), respectively [102]. These complexes did not show any catalytic activity for polymerization, whereas the cationic complexes [Cp$_2$TiMeL] [BPh$_4$] (L = THF, Et$_2$O, MeOPh) (**88**) polymerized ethylene [98]. The protonation of the metallacycles, Cp$_2'$M(CH$_2$SiMe$_2$CH$_2$) (Cp$_2'$ = (C$_5$H$_5$)$_2$, Me$_2$Si(C$_5$H$_4$)$_2$ for M = Ti, Zr; Cp$_2'$ = (C$_5$Me$_5$)$_2$ for M = Zr), with [NEt$_3$H] [BPh$_4$] in THF yielded the corresponding cationic (trimethylsilyl)methyl complexes, [Cp$_2'$M(CH$_2$SiMe$_3$)(THF)][BPh$_4$] (**60**). A THF-free cation complex [Cp$_2^*$Zr(CH$_2$SiMe$_3$)][BPh$_4$] (**89**) was prepared by the reaction of Cp$_2^*$Zr(CH$_2$SiMe$_2$CH$_2$) with [N(n-butyl)$_3$H][BPh$_4$] in toluene. The THF-free

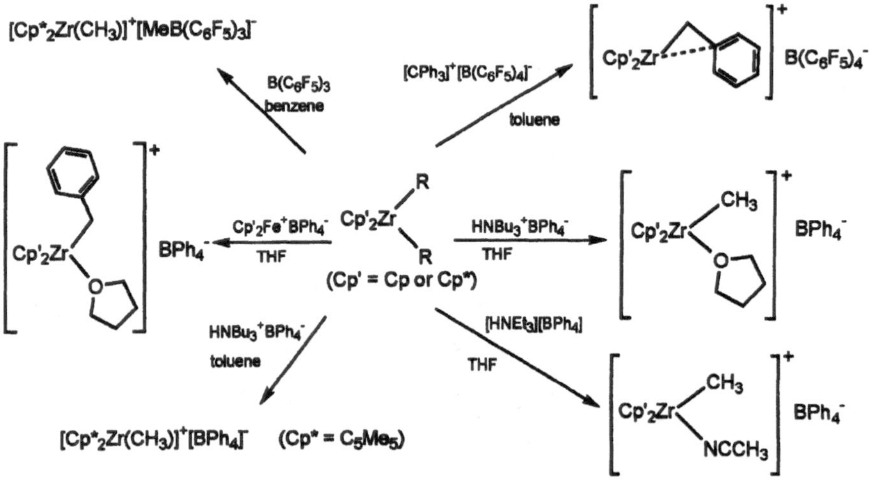

Scheme 2.

Table 5. Catalytic activity of selected cationic metallocene complexes of Group 4 metals for the polymerization of ethylene

Complex	Temperature (°C)	Activity	$M_n/10^3$	M_w/M_n	Ref.
[Cp$_2$ZrMe(thf)][BPh$_4$] (59)	23	12	30	2.5	[76,81]
[Cp$^*_2$ZrMe(tht)][BPh$_4$] (71a)	23	170	21	3.4	[93]
Cp$^*_2$Zr$^+$(m-C$_6$H$_4$)B$^-$(C$_6$H$_4$Me-4)$_4$ (72)	80	375			[94]
[Cp$_2^*$ZrMe][C$_2$B$_9$H$_{12}$] (73)	40	265			[94]
[Cp$_2$ZrMe][(C$_2$B$_9$H$_{11}$)$_2$Co] (74)	60	400	3.5–9.4	3.3–11.6	[54]
[Cp$_2$ZrMe][MeB(C$_6$F$_5$)$_3$] (75)	25	4500	612	2.0	[79]
[C$_5$H$_3$Me$_2$-1,2)$_2$ZrMe][MeB(C$_6$F$_5$)$_3$] (76)	25	6800	367	1.4	[95]
[C$_5$H$_4$SiMe$_3$)$_2$ZrMe][B(C$_6$F$_5$)$_4$] (77)	21	2870	114.6	2.6	[96]
[(C$_2$H$_4$)(Ind)$_2$Zr(CH$_2$Ph)][B(C$_6$F$_5$)$_4$] (78)	60	17000	36	4.0	[97]
[Cp*HfMe(tht)][BPh$_4$] (69)	23	170	11	3.1	[93]
[(C$_5$H$_4$SiMe$_3$)$_2$HfMe][B(C$_6$F$_5$)$_4$] (79)	41	1200	15.2	2.3	[96]
[Cp$_2$TiMe(thf)][BPh$_4$] (80)	25	3.1			[98]
[(Ind)$_2$TiMe][BPh$_4$] (81)	−40	2.96	16	2.0	[99]
[(C$_5$H$_4$SiMe$_3$)$_2$TiMe][B{C$_6$H$_3$(CF$_3$)$_2$}$_4$] (82)	0	35	6.01	2.2	[96]

Table 6. Catalytic activity of selected cationic metallocene complexes of Group 4 metals for the polymerization of propylene

Complex	Temperature (°C)	Activity[a]	$M_n/10^3$[b]	M_w/M_n	Tacticity	Ref.
[(C$_2$H$_4$)(Ind)$_2$ZrMe][B(C$_6$F$_5$)$_4$] (83)	−55	>10000	160		0.963[c]	[100]
[(C$_2$H$_4$)(Ind)$_2$ZrMe][B(C$_6$F$_5$)$_4$] (83)	−20	210000			0.94[c]	[101]
[(C$_2$H$_4$)(Ind)$_2$Zr(CH$_2$Ph)][B(C$_6$F$_5$)$_4$] (78)	23	4690	27	2.1	mmm 0.76	[97]
[Me$_2$C(Cp)(Flu)Zr(CH$_2$Ph)][B(C$_6$F$_5$)$_4$] (84)	23	290	22	2.0	rrrr 0.90	[97]
[(C$_2$H$_4$)(Ind)$_2$ZrMe][MeB(C$_6$F$_5$)$_3$] (85)	−20	700				[101]

[a] Activity: kg(mol of catalyst)$^{-1}$ h^{-1} atm^{-1}.
[b] M_w value.
[c] Weight fraction of PP insoluble in refluxing n-hexane or heptane.

complex had activity for ethylene polymerization, but the THF complex had almost no activity. Thus, the coordination of THF significantly retarded the catalytic activity of cationic alkylmetallocenes [82]. A similar reaction of $Cp_2^*ZrMe_2$ with $[n\text{-}Bu_3NH][B(C_6H_4R\text{-}4)_4]$ (R = H, Me, Et) in toluene afforded also the THF-free zwitterions $Cp_2^*Zr^+(m - C_6H_4)B^-(C_6H_4R - 4)_4$ (72). $Cp_2^*ZrMe_2$ and $(C_5Me_4Et)_2ZrMe_2$ were reacted with a diprotic carborane to give $Cp_2'ZrMe(C_2B_9H_{12})$ (Cp' = Cp*, C_5Me_4Et) (73) in which the $[Cp_2'ZrMe]^+$ cation is bound to the $[C_2B_9H_{12}]^-$ anion solely through a Zr–H–B bond. These complexes have 20–30 times higher activities for the polymerization of ethylene than do the base-coordinated metallocene catalysts such as $[Cp_2ZrMe(thf)]$ $[BPh_4]$ (59) [94]. Base-free cationic titanium alkyls have also been generated. Treatment of a dichloromethane solution of $(Ind)_2TiMe_2$ with $[PhNMe_2H][BPh_4]$ gives a cationic complex $[(Ind)_2TiMe][BPh_4]$ (81). Solutions of this complex polymerize ethylene at between -60 and $+10°C$. The activity shows a maximum at ca. $-20°C$ and gradually decreases on raising temperature. Rapid decrease of the activity was observed above 15°C due to the decomposition of cationic species in dichloromethane [99].

The 16-electron complex, $[Cp_2ZrMe(thf)][BPh_4]$ (59), underwent hydrogenation to the insoluble hydride complex $[Cp_2'Zr(H)(thf)][BPh_4]$ (90) by σ-bond-metathesis with the coordinated H_2 molecule [103]. Isobutylene reversibly inserted itself into the cationic hydride $[Cp_2'Zr(H)(thf)][BPh_4]$ (Cp' = C_5H_4Me) to afford $[Cp_2'Zr(CH_2CHMe_2)(thf)][BPh_4]$ (91) [90]. In contrast 18-electron complexes such as $[Cp_2ZrMeL_2][BPh_4]$ (L = CH_3CN, PMe_3, PMe_2Ph, $PMePh_2$ (92) were not hydrogenated [103]. Thus, the phosphine ligand remarkably stabilizes cationic 16 electron complexes by two-electron donation. In a cationic hafnium-phosphine complex, $[Cp_2^*Hf(CH_2CHMe_2)(PMe_3)][BPh_4]$ (70), α-agostic interaction was observed by NMR spectroscopy, indicating the presence of an electrophilic metal center [90].

The weakly coordinated cationic THT (tetrahydrothiophene) complexes, $[Cp_2^*MMe(THT)][BPh_4]$ (71) (M = Zr (a), Hf (b)) have higher activities for ethylene polymerization compared to THF derivatives. The Zr complex 71a oligomerizes propylene to give oligomers up to C_{24}, while the Hf complex gives only the dimer 4-methyl-1-pentene and the trimer 4,6-dimethyl-l-heptene [93]. The THT complexes 71 oligomerize propylene in N,N-dimethylaniline. At room temperature, complex 71a gave a mixture of C_6 to C_{24} oligomers while 71b derivative selectively afforded one trimer (4,6-dimethyl-l-heptene) [83].

The use of weakly coordinating and fluorinated anions such as $B(C_6H_4F\text{-}4)_4^-$, $B(C_6F_5)_4^-$, and $MeB(C_6F_5)_3^-$ further enhanced the activities of Group 4 cationic complexes for the polymerization of olefins and thereby their activity reached a level comparable to those of MAO-activated metallocene catalysts. Base-free cationic metal alkyl complexes and catalytic studies on them had mainly been concerned with cationic methyl complexes, $[Cp_2M\text{-}Me]^+$. However, their thermal instability restricts the use of such systems at technically useful temperatures. The corresponding thermally more stable benzyl complexes,

[Cp$_2'$Zr(CH$_2$Ph)][B(C$_6$F$_5$)$_4$] (93), where Cp$_2'$ = Cp$_2$, (C$_5$H$_4$SiMe$_3$)$_2$, (C$_2$H$_4$)(Ind)$_2$, (Me$_2$C)(Cp)(Flu), were found to be useful at higher temperatures [97]. Comparing two counter anions, B(C$_6$F$_5$)$_4^-$ and MeB(C$_6$F$_5$)$_3^-$, the former was found to be superior for propylene polymerization [101].

Chien et al. reported isospecific polymerization of propylene when catalyzed by [(C$_2$H$_4$)(Ind)$_2$ZrMe][B(C$_6$F$_5$)$_4$] (83). The activity of the complex increases with decreasing polymerization temperature [100]. Bochmann and his coworkers have reported that the reaction of Cp$_2'$MMe$_2$ (M = Ti, Zr, or Hf; Cp' = C$_5$H$_4$SiMe$_3$) with [CPh$_3$][BR$_4$] (R = C$_6$H$_3$(CF$_3$)$_2$, C$_6$F$_5$) gives base-free cationic [Cp$_2'$MMe]$^+$ catalysts. The ethylene polymerization activity of a series of complexes [Cp$_2'$MMe]$^+$ decreases in the order M = Ti ≪ Hf < Zr. The activities of the Zr and Hf catalysts are comparable with those of Cp$_2$MCl$_2$/MAO systems [for [Cp$_2'$ZrMe][B(C$_6$F$_5$)$_4$], 2870 kg (mol of catalyst)$^{-1}$h^{-1}atm^{-1}] (77) [96].

The reaction of zirconocene dimethyl complexes with B(C$_6$F$_5$)$_3$ gave [L$_2$ZrMe][MeB(C$_6$F$_5$)$_3$] (L = C$_5$H$_5$, C$_5$H$_3$Me$_2$-1,2, C$_5$Me$_5$) (94) [79]. The crystal structure of [(C$_5$H$_3$Me$_2$-1,2)$_2$ZrMe][MeB(C$_6$F$_5$)$_3$] (76) revealed weak coordination of the MeB(C$_6$F$_5$)$_3^-$ anion to the zirconium. These complexes have activities for ethylene polymerization that are comparable to those of typical zirconocene/MAO catalysts [79]. Similarly, the reaction of other zirconocene dimethyl derivatives, rac-Me$_2$Si(Ind)$_2$ZrMe$_2$ and {C$_5$H$_3$(SiMe$_3$)$_2$-1,3}$_2$ZrMe$_2$, with B(C$_6$F$_5$)$_3$ also afforded cationic zirconocene derivatives weakly coordinated by the MeB(C$_6$F$_5$)$_3^-$ anion, while the reaction of dibenzyl complexes, Cp$_2'$Zr(CH$_2$Ph)$_2$ (Cp$_2'$ = rac-Me$_2$Si(Ind)$_2$, {C$_5$H$_3$(SiMe$_3$)$_2$-1,3}$_2$, (C$_5$H$_5$)$_2$, rac-C$_2$H$_4$(Ind)$_2$), leads to [Cp$_2'$ZrCH$_2$Ph][PhCH$_2$B(C$_6$F$_5$)$_3$] (95) which have η^2-coordinated benzyl ligands and noncoordinated [PhCH$_2$B(C$_6$F$_5$)$_3$] counter anions. The structurally similar complexes, [Cp$_2'$ZrCH$_2$Ph][B(C$_6$F$_5$)$_4$] (96), were obtained from the corresponding zirconocene dibenzyl complex with [CPh$_3$][B(C$_6$F$_5$)$_4$], which are stable in toluene at 60°C. The complexes are highly active catalysts for the polymerization of ethylene and propylene [80].

The oxidative reaction of Cp$_2$ZrR$_2$ with Ag[BPh$_4$] of [Cp$_2$Fe][BPh$_4$] in CH$_3$CN afforded [Cp$_2$Zr(R)(CH$_3$CN)][BPh$_4$] (R = H, Ph, CH$_3$, η^2-CH$_2$Ph) (97) [104, 105]. Insertion of the coordinated acetonitrile of complexes 97 to the zirconium-alkyl bond yielded [Cp$_2$Zr{N = C(R)(CH$_3$)}(CH$_3$CN)] [BPh$_4$] (98) [105]. The oxidative reaction of Cp$_2'$Zr(CH$_2$Ph)$_2$ (Cp' = C$_5$H$_4$Me) with [Cp$_2'$Fe][BPh$_4$] affords a normal η^1-benzyl complex [Cp$_2'$Zr(CH$_2$Ph)(thf)][BPh$_4$] (99), which undergoes hydrogenolysis to produce a cationic hydride complex [Cp$_2'$Zr(H)(thf)][BPh$_4$] (90). This hydride complex reacts with allene to give an η^3-allyl complex [Cp$_2'$Zr(η^3-C$_3$H$_5$)(thf)][BPh$_4$] (100). These complexes oligomerize ethylene in CH$_2$Cl$_2$ [106]. Similarly rac-(C$_2$H$_4$)(Ind)$_2$Zr(CH$_2$Ph)$_2$ reacts with [Cp$_2'$Fe][BPh$_4$] to yield [rac-(C$_2$H$_4$)(Ind)$_2$Zr(CH$_2$Ph)(thf)][BPh$_4$] (101), which undergoes ligand substitution in CH$_3$CN to give an η^2-benzyl complex [rac-(C$_2$H$_4$)(Ind)$_2$Zr(η^2-CH$_2$Ph)(CH$_3$CN)][BPh$_4$] (102) [107].

The reaction of $Ag[CB_{11}H_{12}]$ with Cp_2ZrR_2 ($R = CH_2Ph$, Me) yields $Cp_2ZrR(\eta^1\text{-}CB_{11}H_{12})$ (**103**) where $CB_{11}H_{12}^-$ is another weakly coordinating anion. The benzyl complex polymerizes ethylene in toluene above 0°C. [108]. The methyl complex oligomerizes propylene primarily to 2-methyl-1-pentene and 2,4-dimethyl-1-heptene. $Cp^*ZrMe_2(\eta^3\text{-}CB_{11}H_{12})$ (**104**) is prepared from the reaction of Cp^*ZrMe_3 with $Ag[CB_{11}H_{12}]$. It is unreactive with propylene, 2-butyne, and styrene [109].

Anions of the type $[M(C_2B_9H_{11})_2]^-$ (M = Fe, Cp, Ni) were also used as noncoordinating anions with $[Cp_2ZrMe]^+$, which are active for the polymerization and copolymerization of ethylene and α-olefins in non-polar solvents such as toluene and hexane [54]. By using the same anions, cationic actinide complexes have also been prepared [110].

3.2 Neutral Group 3 Metallocene Complexes as Catalysts of Polymerization

Neutral isoelectronic Group 3 metallocene complexes such as Cp_2^*MR (M = Sc, Y, and lanthanide metals) (Table 7) are *14 electron species* and show an isolobal analogy to cationic Group 4 metallocene species as shown schematically in Chart 1. They are found to be highly active catalysts for the polymerization of ethylene [56, 111–116]. The activity depends upon the kind of metal center; i.e. the activity is in the order of La $\geqslant$ Nd $\gg$ Lu [56, 117]. The early lanthanide metallocene hydrides showed higher activities than those of homogeneous Group 4 metal based catalysts at the initial stage of the polymerization, although they showed decreasing activities with increasing reaction time. The polymerization of ethylene catalyzed by a hydride complex of lutetium, Cp_2^*LuH (**105**), afforded polyethylene with rather narrow polydispersity, $M_w/M_n = 1.37$ [56]. Cationic complexes of actinides such as $[Cp_2^*ThMe]$ $[B(C_6F_5)_4]$ (**106**) were found to be active catalysts for the polymerization of ethylene [118].

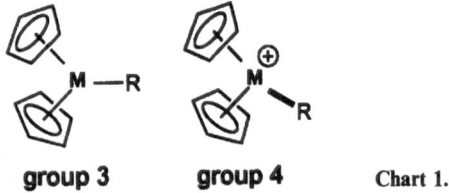

group 3   **group 4**   Chart 1.

On the other hand, the reaction of Cp_2^*LnR with propylene did not afford any polymers but rather an allyl complex, $Cp_2^*Ln(\eta^3\text{-allyl})$, via a σ-bond meta-thesis reaction [56, 117]. One molecule of propylene can insert itself into the Lu–Me bond of Cp_2^*LuMe to give the corresponding isobutylene complex. The successive insertion of propylene is 1000-fold slower than the first insertion [57]. The gas-phase reaction of $Sc(CH_3)_2^+$ with propylene also produces a

Table 7. Catalytic activity of metallocene complexes of Group 3 metals

Complex	Temperature (°C)	Activity[a]	$M_n/10^3$	M_w/M_n	Ref.
Cp*LaH (40)	25	182000	680	2.03	[56]
Cp*$_2$NdH (107)	25	137000	590	1.81	[56]
Cp*$_2$LuH (105)	25	10000	96	1.37	[56]
Cp*$_2$LuMe (41)	50–80	6900	–	–	[57]
Me$_2$Si(C$_5$H$_2$-2-SiMe$_3$-4-SiMe$_2$Bu^t)$_2$YH (108)	25	58.4	82	4.18	[119]
{Cp*Y(OC$_6$H$_3$Bu^t-2,6)$_2$(μ-H)}$_2$ (109)	25	0.34	12	10.8	[120]
[Cp*$_2$ThMe][BPh$_4$] (110)	25	1.1			[121, 122]
[Cp*$_2$ThMe][B(C$_6$F$_5$)$_4$] (106)	25	360			[122]

[a] Activity: kg(mol of catalyst)$^{-1}$ h^{-1} atm^{-1}

σ-bond metathesis products, ScMe(allyl) and Sc(allyl)$_2$ [123]. In the case of yttrium, [(C$_5$H$_4$Me)$_2$YH(thf)]$_2$ reacts with ethylene and propylene to form (C$_5$H$_4$Me)$_2$YR(thf) (R = ethyl, n-propyl) without polymerization [124].

A scandium complex, Cp$_2^*$ScH, also polymerizes ethylene, but does not polymerize propylene and isobutene [125]. On the other hand, a linked amidocyclopentadienyl complex [{Me$_2$Si(η^5-C$_5$Me$_4$)(η^1-NCMe$_3$)}Sc(H)(PMe$_3$)]$_2$ slowly polymerizes propylene, 1-butene, and 1-pentene to yield atactic polymers with low molecular weight (M_n = 3000–7000) [126, 115]. A chiral, C_2-symmetric *ansa*-metallocene complex of yttrium, [*rac*-Me$_2$Si(C$_5$H$_2$SiMe$_3$-2-Bu^t-4)$_2$YH]$_2$, polymerizes propylene, 1-butene, 1-pentene, and 1-hexene slowly over a period of several days at 25°C to afford isotactic polymers with modest molecular weight [114].

Marks et al. reported the co-polymerization of ethylene and 1-hexene by using *ansa*-type complexes of lanthanide metals [127]. Recently, bulky alkyl substituted *ansa*-type metallocene complexes of yttrium have been reported to exhibit high activity for the polymerization of 1-hexane. [114, 119, 128]

3.3 Gas-Phase Reaction of Cationic d^0 Alkyl Compounds of Group 4 Metals

Remarkably high reactivity of cationic alkyl complexes of Group 4 metals with 1-alkenes has been observed in gas-phase reactions [129]. Typical ionic species such as TiCl$_2$Me$^+$ react with ethylene, and the insertion followed by H$_2$ elimination gives rise to a cationic allyl complex TiCl$_2$C$_3$H$_5^+$, which does not react further with ethylene.

The gas-phase reaction of cationic zirconocene species, ZrMeCp$_2^+$, with alkenes and alkynes was reported to involve two major reaction sequences, which are the migratory insertion of these unsaturated hydrocarbons into the Zr–Me bond (Eq. 3) and the activation of the C–H bond via σ-bonds metathesis rather than β-hydrogen shift/alkene elimination (Eq. 4) [130, 131]. The insertion in the gas-phase closely parallels the solution chemistry of Zr(R)Cp$_2^+$ and other isoelectronic complexes. Thus, the results derived from calculations based on this gas-phase reactivity should be correlated directly to the solution reactivity (vide infra).

$$\text{Cp}_2\text{ZrCH}_3^+ + \text{H}_2\text{C=CHR} \longrightarrow \left[\text{Cp}_2\text{Zr}\underset{\overset{\displaystyle C}{\underset{\displaystyle H_2}{}}}{\overset{\overset{\displaystyle H_3}{\displaystyle C}}{}}\text{CHR} \right]^{+\ddagger} \longrightarrow \text{Cp}_2\text{Zr}^+\underset{\overset{\displaystyle C}{\underset{\displaystyle H_2}{}}}{\overset{\displaystyle CH_3}{}}\text{CHR} \qquad (3)$$

$$\text{Cp}_2\text{ZrCD}_3^+ + \text{RH} \longrightarrow \left[\text{Cp}_2\text{Zr}\underset{\overset{\displaystyle}{\underset{\displaystyle R}{}}}{\overset{\overset{\displaystyle D_3}{\displaystyle C}}{}}\text{H} \right]^{+\ddagger} \longrightarrow \text{Cp}_2\text{ZrR}^+ + \text{CD}_3\text{H} \qquad (4)$$

The Group 3 metal cationic species $ScMe_2^+$ similarly reacts with ethylene to give $Sc(Me)(allyl)^+$ after H_2 elimination. This species reacts further with ethylene to form the dehydrogenated product $Sc(allyl)_2^+$ [132]. The reaction of these d^0 species with propylene gave allyl complexes via a σ-bond metathesis activating the allylic C–H bond. Although the insertion reaction is normally observed for cationic d^0 metal systems, d^n species mainly undergo σ-bond metathesis reactions rather than the migratory insertion in the gas-phase reaction of cationic metal-methyl MMe^+ (where M = Fe, Y and so on) with alkenes [133]. Thus, this evidence collectively suggests that the coordinatively unsaturated d^0 species is important for the insertion of alkene into an M–C bond.

3.4 Propagation

In the propagation process of Ziegler-Natta polymerization, the insertion of olefin into a metal-carbon bond is the most important basic step, but many questions concerning to this process remained unanswered for a long time.

The most famous mechanism, namely *Cossee's mechanism*, in which the alkene inserts itself directly into the metal-carbon bond (Eq. 5), has been proposed, based on the kinetic study [134–136], This mechanism involves the intermediacy of ethylene coordinated to a metal-alkyl center and the following insertion of ethylene into the metal-carbon bond via a four-centered transition state. The olefin coordination to such a catalytically active metal center in this intermediate must be weak so that the olefin can readily insert itself into the M–C bond without forming any meta-stable intermediate. Similar alkyl-olefin complexes such as $Cp_2NbR(\eta^2$-ethylene) have been easily isolated and found not to be the active catalyst precursor of polymerization [31–33, 137]. In support of this, theoretical calculations recently showed the presence of a weakly ethylene-coordinated intermediate (vide infra) [12, 13]. The stereochemistry of ethylene insertion was definitely shown to be *cis* by the evidence that the polymerization of *cis*- and *trans*-dideutero-ethylene afforded stereoselectively deuterated polyethylenes [138].

$$L_nM-CH_2\textcircled{P} \longrightarrow L_nM-CH_2CH_2CH_2\textcircled{P}$$
$$\uparrow$$
$$CH_2{=}CH_2$$

(5)

As the second mechanism, Green and Rooney proposed the '*metallacycle mechanism*' in which metal-carbene and metallacyclobutane are the key intermediates, based on the α-hydrogen elimination from an alkyl-metal bond as shown in Eq. (6) [139]. A theoretical study suggests that this process will occur quite easily [140]. An experimental support for this mechanism has been obtained in the research on some alkylidene complexes of tantalum. A tantalum alkylidene hydride complex, $Ta(=CHCMe_3)(H)L_3I_2$, (where L = PMe_3) reacts with ethylene to give the ethylene-inserted compound $Ta[(=CH(CH_2CH_2)_n$

CMe_3](H)L_3I_2, whose protonolysis afforded polyethylene [141].

$$L_nM-CH_2P \longrightarrow L_n\overset{\overset{\displaystyle H}{|}}{M}=CHP \underset{CH_2=CH_2}{\longrightarrow} L_n\overset{\overset{\displaystyle H}{|}}{M}\overset{\displaystyle -CHP}{\underset{\displaystyle H_2C-CH_2}{|}} \qquad (6)$$

$$\longrightarrow L_nM-CH_2CH_2CH_2P$$

Here we discuss more details of the Cossee mechanism using recent evidence obtained by modern organometallic chemistry, together with relevant results of theoretical calculations.

The Cossee mechanism has been demonstrated by direct observation of organometallic complexes where a C = C bond inserts itself into an M–C bond as shown in Eq. (7)–(9). A labeling experiment on a cationic platinum complex **111** indicated the reversible insertion of the coordinated alkene into the Pt–C bond as shown in Eq. (7) [142].

$$\qquad (7)$$

111   **111**

The direct insertion of propylene into the Lu–Me bond of $Cp_2^*LuMe(OEt_2)$ (**112**) was reported to give isobutyl (**113**) and 2,4-dimethylpentyl (**114**) complexes (Eq. 8) [57, 143].

$$\qquad (8)$$

112   **41**   **113**   **114**

By performing excellent model reactions [144], Grubbs and his co-workers demonstrated direct olefin insertion into an M–C bond. Thus, complex **115** was treated with $AlEtCl_2$ to give complex **116**, whose decomposition afforded methylcyclopentane. Under the same conditions, the polymerization of ethylene took place. In this way, the insertion of α-olefins into a Ti–C single bond in a model Ziegler-Natta catalyst system was directly observed (Eq. 9).

$$\qquad (9)$$

115   **116**

As a model for the insertion process in the polymerization of ethylene, the reaction of Cp_2^*ScMe with 2-butyne was investigated. The reaction was revealed to have a relatively small enthalpy of activation and a very large negative entropy of activation; a highly ordered four-centered transition state (**117**) was proposed [111, 112].

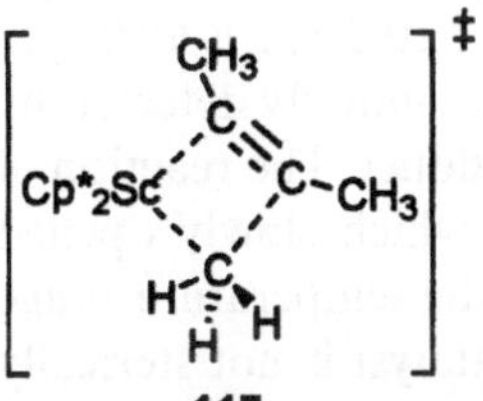

117

The importance of α-*agostic interaction* during the propagation process has been demonstrated by the labeling experiments of Bercaw's and Britzinger's groups independently. Scheme III summarizes the results reported by Bercaw's group [145] and the excellent work by Brintinger's group is discussed in Scheme V (vide infra). The propensity of α-agostic interaction with the α-H rather than the α-D is shown.

The presence of α-agostic interaction in alkyl complexes of early transition metals and actinide metals has recently been confirmed. Thus, a neutron diffraction study of $Cp_2^*Th(CH_2CMe_3)_2$ clearly revealed this interaction [146]. The presence of one of the α-hydrogens on the two neopentyl groups quite near to the metal (2.597(9) and 2.648(9) Å) indicates the agostic interaction. MO calculations of $Cp_2Th(C_2H_5)_2$ indicated that not only a α-H but also a β-H can participate in agostic interactions without losing Th–C bond strength [147, 148].

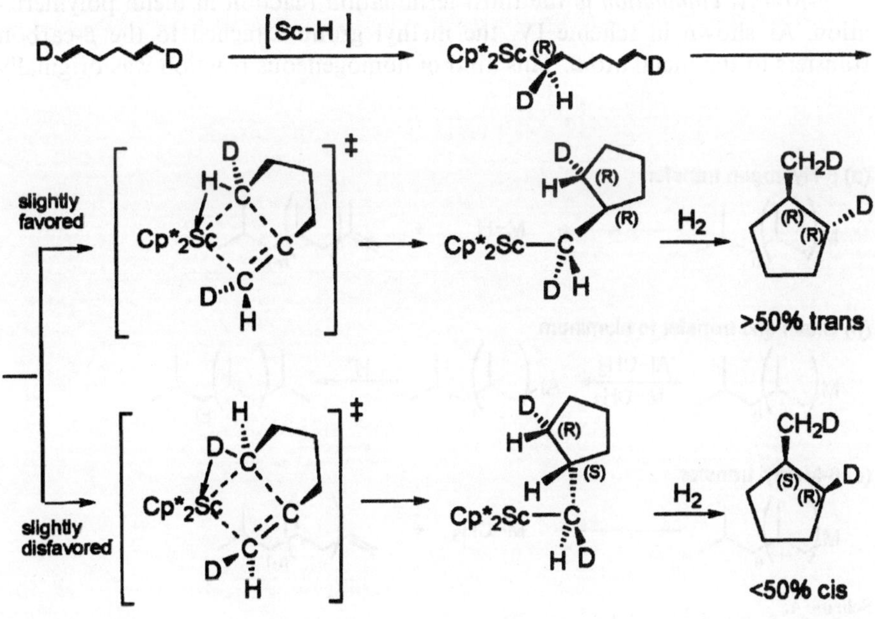

Scheme 3.

3.5 Termination Reactions: β-Hydrogen Elimination, Alkyl Chain End Transfer, and β-Methyl Elimination

Termination reaction of olefin polymerization can be classified into three categories (Scheme IV). The polypropylene prepared by Kaminsky's catalyst system has the terminal C=C double bond, being terminated via *β-hydrogen elimination* [149, 150]. The resulting Cp_2ZrH^+ species was initially detected by XPS [69]. Recently, this species was isolated as a THF adduct. The reaction of $[Cp_2ZrMe(THF)]^+$ with H_2 afforded $[Cp_2ZrH(thf)]^+$, which slowly opened the THF ring to yield $[Cp_2Zr(OBu)(THF)]^+$ [151]. In the temperature range 0–80°C, β-hydrogen elimination occurs only when the catalyst is not sterically crowded. Thus the polymerization using catalyst precursors such as Cp_2ZrCl_2, Cp_2HfCl_2, *rac*-$(C_2H_4)(IndH_4)_2ZrCl_2$, and $Me_2Si(Ind)_2ZrCl_2$ has mostly been accompanied by β-hydrogen transfer reaction [150–154].

The second termination reaction is *alkyl chain end transfer* from the active species to aluminium [155]. This termination becomes major one at lower temperatures in the catalyst systems activated by MAO. ^1H and ^{13}C NMR analysis of the polymer obtained by the cyclopolymerization of 1,5-hexadiene, catalyzed by $Cp_2^*ZrCl_2/MAO$, afforded signals due to methylenecyclopentane, cyclopentane, and methylcyclopentane end groups upon acidic hydrolysis, indicating that chain transfer occurs both by β-hydrogen elimination and chain transfer to aluminium in the ratio of 2:8, and the latter process is predominant when the polymerization is carried out at $-25°C$ [156]. The values of rate constants for Cp_2ZrCl_2/MAO at 70°C are reported to be: $k_p = 168$-1670 $(Ms)^{-1}$, $k_{tr}^{al} = 0.021 - 0.81 s^{-1}$, and $k_{tr}^{\beta} = 0.28 s^{-1}$ [155].

β-Methyl elimination is the third termination reaction in olefin polymerization. As shown in scheme IV, the methyl group attached to the β-carbon transfers to the metal atom. This kind of homogeneous reaction was originally

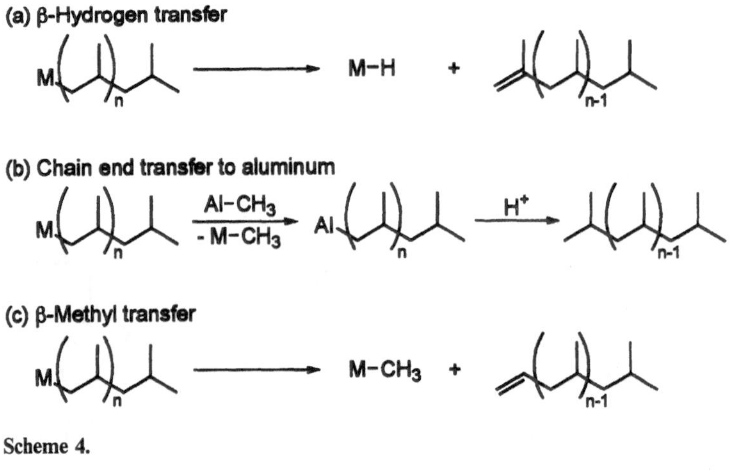

(a) β-Hydrogen transfer

(b) Chain end transfer to aluminum

(c) β-Methyl transfer

Scheme 4.

observed for the reaction of Cp_2^*LuMe with propylene [57,157]. The β-methyl transfer reaction occurs in the case of oligomerization of propylene by the catalyst system based on Cp_2^*Zr, where Cp^* is the pentamethylcyclopentadienyl ligand. Polymerization using $Cp_2^*ZrCl_2$/MAO proceeds with 91% β-methyl transfer, 8% β-hydrogen transfer, and 1% chain transfer to aluminium at 50°C, while the chain transfer to aluminum occurs only at $-40°C$ [152]. Polymerization by a MAO-free system such as $[Cp_2^*ZrMe(THT)]^+[BPh_4]^-$ was terminated almost exclusively by β-methyl transfer at wide range of temperatures (5–95°C) [83]. Such a termination reaction was also observed in the gas-phase reaction of Cp_2ZrMe^+ with isobutene [131]. Competitive β-H and β-Me eliminations were detected in the polymerization of isobutylene, catalyzed by $OpSc(H)(PMe_3)$ $(Op = Me_2Si(C_5Me_4)_2)$ [116].

4 Stereoselective Polymerization of α-Olefins

4.1 Isospecific Polymerization of α-Olefins

The initials Kaminsky-Sinn systems based on Cp_2MX_2/MAO (M = Ti, Zr) afforded only atactic polypropylene. In 1984, Ewen found that the Cp_2TiPh_2/ MAO system became the catalyst for the isospecific polymerization of propylene (mmmm $\approx$ 50%) on cooling down the catalyst system below $-45°C$ [65]. In 1985, Brinzinger designed the rational and effective bridged-metallocenes catalyst system that affords the high isospecificity in α-olefin polymerization. Here we shall briefly review the isospecific polymerization of α-olefin by using bridged- and unbridged-metallocene catalyst systems.

4.1.1 Isospecific Polymerization by Ansa-Type (Bridged) Metallocenes

In 1984, Ewen reported that a mixture of *meso*- and *racemo*-isomers of $(C_2H_4)(Ind)_2TiCl_2$ (50) upon activation with MAO afforded a mixture of isotactic PP and atactic PP [65]. For the first time Brinzinger used a chiral C_2-symmetrical ethylene-bridged metallocene, rac-$(C_2H_4)(IndH_4)_2ZrCl_2$ (29), as catalyst precursor for highly isotactic polymerization of 1-alkenes [59]. This breakthrough was followed by active and widespread research using Group 4 metallocene catalysts with a variety of bridging units. Some representative examples of isospecific polymerization are shown in Table 8 [59]. The isolated isotactic polypropylene was found to have stereoerrors of type 1 in Scheme V (*the enantiomorphic site control*), which is in sharp contrast to the unbridged metallocene catalysts that gave type 2 isotactic polymers in Scheme V through *the chain end control*.

A C_1-symmetric ethylidene-bridged titanium complex, rac-MeCH(Ind) (C_5Me_4)TiCl$_2$ (**51**) showed rather low isospecificity [66]. The polymerization of propylene using rac-MeCH(Ind)(C_5Me_4)TiX$_2$/MAO (X = Cl, Me) gives a stereoblock polymer comprising alternating sequences of stereoregular, crystallizable and stereoirregular, amorphous polypropylene, which is a thermoplastic elastomer [66, 67, 158]. The ethylene-bridged hafnium complexes, (C_2H_4)(Ind)$_2$HfCl$_2$ (**53**) and (C_2H_4)(IndH$_4$)$_2$HfCl$_2$ (**54**), upon activation by MAO became also the catalysts for the isospecific polymerization of propylene [45].

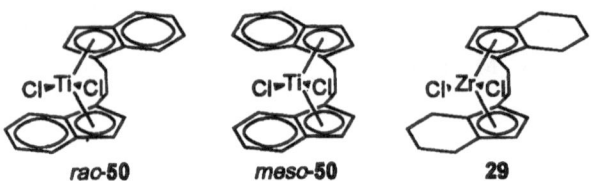

rac-**50**   meso-**50**   **29**

The stereospecificity of homogeneous metallocene catalysts for propylene polymerization is sensitive not only to steric effects but also to electronic effects. With increasing electron density at the metal center, the stereospecificity of the catalyst decreases [49]. The isospecificity of (C_2H_4)(Ind)$_2$ZrCl$_2$/MAO for the polymerization of propylene decreases with increasing polymerization temperature and decreasing [Al]/[Zr] ratio. The polypropylene polymerized by (C_2H_4)(Ind)$_2$ZrCl$_2$/MAO at $T_p = 30°C$ and [Al]/[Zr] = 3200 has [mmmm] = 0.90 and $T_m = 132.3°C$. The PP obtained at $T_p = -15°C$ and [Al]/[Zr] = 2700 has [mmmm] = 0.98 and $T_m = 152.4°C$ [165]. The end group of isotactic PP polymerized by (C_2H_4)(Ind)$_2$TiX$_2$ (X = Cl, Me)/^{13}C-enriched organoaluminium compounds were confirmed to be controlled by the enantiomorphic site control of the chiral catalyst metal center [166]. Investigation on the regio-irregularity of the polypropylenes obtained by rac-(C_2H_4)(Ind)$_2$ZrCl$_2$ and rac-(C_2H_4)(IndH$_4$)$_2$ZrCl$_2$ resulted in the same conclusion [153]. Recently, an ansa-type yttrium without any co-catalyst afforded high isospecificity ($M_n = 4200$, $M_w/M_n = 2.32$, mmmm 97% for propylene) in the polymerization of 1-alkenes [114]. This single component catalyst is expected to be the most convenient system for the mechanistic study.

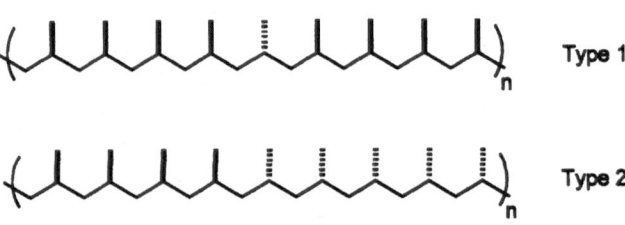

Type 1

Type 2

Scheme 5.

Table 8. Isotactic poly(propylene) polymerized by bridged- and unbridged-metallocene catalysts

Metal	Ligand	Co-catalyst	$M_n/10^3$	Tacticity (%)	Ref.
Ansa-metallocenes					
Zr	(C$_2$H$_4$)(Ind)$_2$ (29)	MAO	6.3	63(mmmm)	[97]
	rac-(C$_2$H$_4$)(IndH$_4$)$_2$ (29)	MAO	41–280	95(mm)	[59]
	Me$_2$Si(Ind)$_2$ (35)	MAO	20	96.2(mm)	[53]
	Me$_2$Si(C$_5$H$_2$Me$_3$-2,3,5)(C$_5$HMe$_3$-2',4'5') (43)	MAO	67	97.7(mmmm)	[60]
	Me$_2$Si(C$_5$H$_2$Bu^t-3-Me-5)(C$_5$H$_3$Me-2'-Bu^t-4') (118)	MAO	3.7	94(mmmm)	[61]
	Me$_2$Si(Benz[e]indenyl)$_2$ (119)	MAO	28	93.9(mmmm)	[159]
	Me$_2$Si(2-Me-Benz[e]indenyl)$_2$ (120)	MAO	261	96.2(mmmm)	[159]
	Me$_2$Si(IndMe-2-Ph-4)$_2$ (121)	MAO	729^a	95.2(mmmm)	[160]
	Me$_2$Si(IndMe-2-Naph-4)$_2$ (122)	MAO	920^a	99.1(mmmm)	[160]
Hf	(C$_2$H$_4$)(Ind)$_2$ (53)	MAO	≥ 329	95–99^b	[45]
	(C$_2$H$_4$)(IndH$_4$)$_2$ (54)	MAO	68	95–99^b	[45]
	Me$_2$Si(C$_5$H$_2$Me$_3$-2,3,5)(C$_5$HMe$_3$-2',4',5') (123)	MAO	110	98.7(mmmm)	[60]
Ti	(C$_2$H$_4$)(Ind)$_2$ (50) (56% rac 44% meso)	MAO	97	54(mm)	[65]
Y	rac-MeCH(Ind)(C$_5$Me$_4$) (51)	MAO	67	40(mmmm)	[66]
	rac-Me$_2$Si(C$_5$H$_2$SiMe$_3$-2-Bu^t-4)$_2$ (124)	–	4.2	97.0(mmmm)	[114]
Unbridged metallocenes					
Zr	C$_5$H$_4$CHMePh (125)	MAO	10^c	60	[161]
	1-methylfluorenyl (126)	MAO	–	83	[162]
	1-neoisomethylindenyl (127)	MAO	100^c	77	[163]
	3-α-cholestanylindenyl (128)	MAO	470^c	80	[164]

^{a} M_w value.
^{b} Mole percentage of units with the same relative configuration.
^{c} M_η value.

The C_2-symmetric silylene bridged metallocenes such as rac-Me$_2$Si(Ind)$_2$ ZrCl$_2$ (35) [53], Me$_2$Si(C$_5$HMe$_3$-2,3,5)(C$_5$HMe$_3$-2',4',5')ZrCl$_2$ (43) [60], Me$_2$Si (C$_5$H$_2$Bu^t-3-Me-5)(C$_5$H$_3$Me-2'-Bu^t-4')ZrCl$_2$ (118) [61], Me$_2$Si(Benz[e]indenyl)$_2$ ZrCl$_2$ (119) [63, 159, 160], and Me$_2$Si(2-Me-Benz[e]indenyl)$_2$ZrCl$_2$ (120) [63, 159, 160] were used as the catalyst precursors for the isospecific polymerization of propylene. The catalyst 119/MAO was shown to polymerize propylene with high activity producing rather low molecular weight polymer (at 4 atm, $M_n = 2.8 \times 10^4$). On the other hand, 120/MAO yielded a polymer of higher molecular weight (at 4 atm, $M_n = 2.2 \times 10^5$). Thus, the 2-methyl substitution of the $ansa$-cyclopentadienyl derivatives effectively enhances the molecular weight of the resulting polymer [159].

43   118   119   120

The use of dimethylbis(2-methyl-4-(1-aryl)indenyl)silane (121 and 122) as a ligand was found to give the poly(propylene) with high stereoregularity and high molecular weight [160]. The highest isospecificity among the metallocene catalysts was achieved by the system of 122.

121   122

C_2-symmetric doubly bridged zirconocene complexes such as (Me$_2$Si)$_2$(1,2-C$_5$HMe$_2$-3,4)$_2$ZrCl$_2$ (123) and (Me$_2$Si)$_2$(1,2-IndH$_4$)$_2$ZrCl$_2$ (124) were reported to have been converted into active catalysts for the polymerization of propylene, yielding polymers with relatively low isotacticities (mmmm = 38 and 80%, respectively) [167].

123    **124**

For the isotactic polymerization of propylene by using chiral *ansa*-zirco-nocene, the presence of α-*agostic interaction* in the propagation process and the origin of occasional stereo-errors has been demonstrated by a deuterium labeling study [168]. In the polymerization of (E)- and (Z)-propene-*1-d* by *rac*-C$_2$H$_4$(indH$_4$)$_2$ZrCl$_2$/MAO, *rac*-Me$_2$Si(C$_5$H$_2$Me$_2$-2,4)$_2$ZrCl$_2$/MAO, or *rac*-Me$_2$Si(C$_5$H$_2$Me-2-Bu^t-4)$_2$ZrCl$_2$/MAO, the polymers derived from (E)-propene-*1-d* have mean degree of polymerization (P_N) values that are 1.3 times higher than those obtained from (Z)-propene-*1-d*. The mean degree of poly-merization can be given by the ratio of the rates of polymer growth (v_P) and chain termination (v_T), $P_N = v_P/v_T$. The *re*-facial insertion, (E)-propene-*1-d* will give α-(R)-ZrCHDR and (Z)-propene-*1-d* will give α-(S)-ZrCHDR (Scheme VI). As chain growth in this system is terminated mainly by β-H transfer, the configuration of α-CHD would not affect v_T. On the other hand, v_P can depend on the α-CHD configuration if one of the α-hydrogen atoms is in contact with the Zr center. Thus, α-H agostic interactions are operative also in the isospecific propylene polymerization with chiral *ansa*-metallocene catalysts. The D atom coupled ^{13}CNMR study of the resulting polypropylene indicated that the stereo-errors produced in these catalyst systems arise from chain-end isomeriz-ation rather than from errors in the enantiofacial orientation of the inserting olefin.

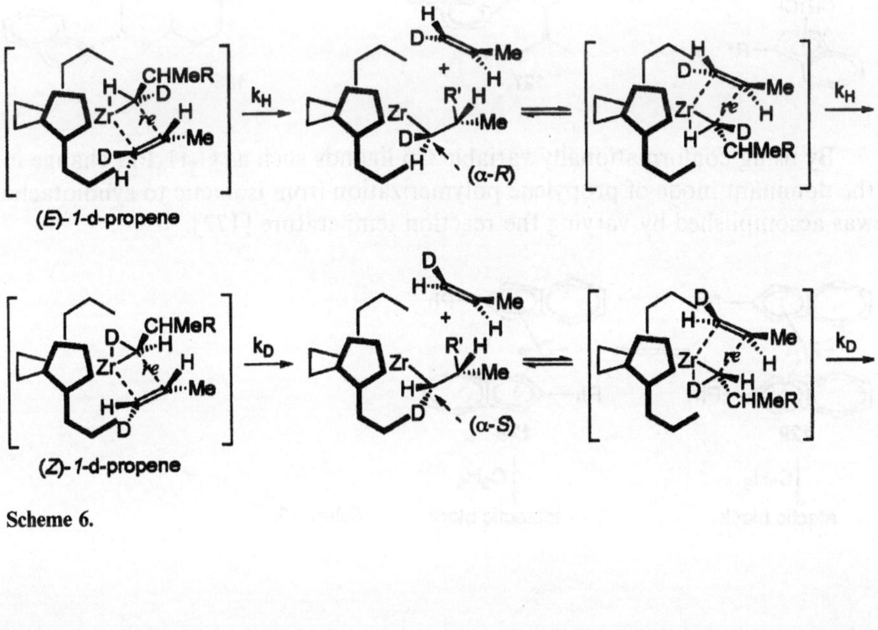

Scheme 6.

4.1.2 Isospecific Polymerization by Unbridged Metallocenes

The isotacticities and activities achieved with nonbridged metallocene catalyst precursors were low. Partially isotactic polypropylene has been obtained by using a catalyst system of unbridged (non-*ansa* type) metallocenes at low temperatures [65]. A chiral zirconocene complex such as *rac*-ZrCl$_2$(C$_5$H$_4$CHMePh)$_2$ (125) is the catalyst component for the isospecific polymerization of propylene (mmmm 0.60, 35% of type 1 and 65% of type 2 in Scheme V) [161]. More bulky metallocene such as bis(1-methylfluorenyl)zirconium dichloride (126) together with MAO polymerized propylene to isotactic polypropylene in a temperature range between 40 and 70°C [162].

To examine the effect of bulky substituents on the Cp ring, hydroboration of (C$_5$H$_4$CH=CMe$_2$)$_2$ZrCl$_2$ with 9-borabicyclo[3.3.1]nonane was utilized to give a mixture of *rac*- and *meso*-{C$_5$H$_4$CH(BC$_8$H$_{14}$)Pr^i}$_2$ZrCl$_2$, which was separated by fractional crystallization. Polymerization of propylene using *meso*-{C$_5$H$_4$CH(BC$_8$H$_{14}$)Pr^i}$_2$ZrCl$_2$/MAO at -50°C afforded isotactic polypropylene by chain end control ($\sigma \approx 0.7$), while the use of *rac*-{C$_5$H$_4$CH(BC$_8$H$_{14}$)Pr^i}$_2$ZrCl$_2$ as a catalyst precursor *it*-PP by enantiomorphic site control ($\alpha \approx 0.95$) [169]. Similar precursors, *rac*-(C$_5$H$_4$CHR^{1}R^2)$_2$ZrCl$_2$ (R^1, R^2 = Me,Cy; Me,Ph; CH$_2$B(C$_8$H$_{14}$),Cy; CH$_2$B(C$_8$H$_{14}$),Ph), were also examined [170]. Polypropylene obtained with Zr complexes containing bulky chiral side chains, (R*Ind)$_2$ZrCl$_2$/MAO, was also highly isotactic (mmmm = 77% for R*Ind = neoisomenthylindenyl (127), mmmm = 80% for R*Ind = 3-α-cholestanylindenyl (128)) [163, 164, 171].

127   128

By using conformationally variable Cp ligands such as C$_5$H$_4$Pr^i, change in the dominant mode of propylene polymerization from isotactic to syndiotactic was accomplished by varying the reaction temperature [172].

129   129

atactic block   isotactic block   Scheme 7.

Waymouth and coworkers reported a unique system where the unbridged bis(2-phenylindenyl)zirconium-based catalysts (129) gave elastomeric, isotactic-atactic stereoblock polypropylene, controlled by rotation of the 2-phenylindenyl as shown in Scheme VII [173].

4.2 Syndiospecific Polymerization of α-Olefins

4.2.1 Ansa-Type Metallocene Catalyst

Highly syndiotactic poly(alkene) has been obtained by Ewen and his co-workers using Me$_2$C(Cp)(Flu)MCl$_2$ (M = Zr (48), Hf (130)) [64]. The stereochemistry during propagation is the result of the site control in the coordination of alkene to a metal center. It has been proposed that the fluorenyl group causes direct prochiral face control of the coordination of alkene to active site by the direct steric interaction. Recently, however, indirect control by the fluorenyl group has been suggested by detailed conformational analysis using molecular mechanics [174, 175]. According to the study, the direct interactions of the Me$_2$C(Cp)(Flu) ligand with the growing chain determine its chiral conformation, which, in turn, discriminates between the two prochiral faces of the propylene monomer (Scheme VIII).

A great variety of modifications to the steric properties, based on Ewen's work, has been carried out in order to control the stereoselectivity; some representative examples are summarized in Table 9. Using modified Me$_2$C(Cp)(Flu)MCl$_2$ (M = Zr, Hf), high syndiospecificity has been achieved 134–136, as reported by Fina, Hoechst and Mitsui Toatsu Chemicals, inc.

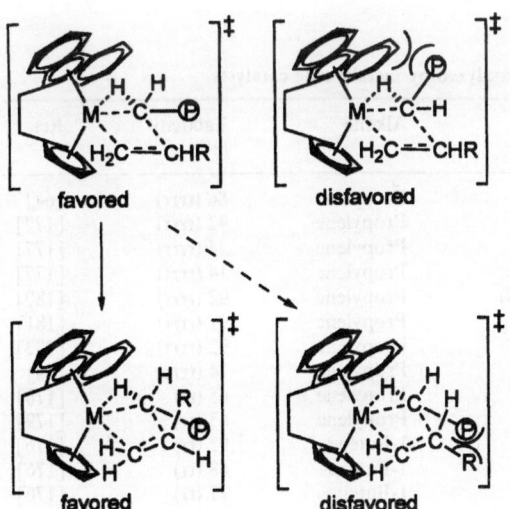

Scheme 8.

4.2.2 Non-bridged Metallocene Catalysts

In contrast to the case of Cp_2ZrX_2/MAO giving atactic poly(alkene)s, $Cp_2^*MCl_2/MAO$, M = Zr (139) and Hf (140), are the catalyst precursors of the syndiotactic polymerization of 1-butene and propylene [176]. Triad distribution indicated that this is chain-end controlled syndiospecific polymerization. The syndiospecificity is attributed to the increase of steric encumbrance around the metal center. Thus, $Cp_2^*HfX_2$ is the most effective syndiospecific catalyst component in this system.

Table 9. Syndiospecific polymerization catalyzed by metallocene catalysts

Metal	Ligand	Alkene	Tacticity (%)	Ref.
Zr	$Me_2C(Cp)(Flu)$ (48)	Propylene	86 (rrrr)	[64]
	$Me_2C(Cp)(Cpphen)$ (131)	Propylene	72 (rrrr)	[177]
	$Me_2Si(Cp)(Cpphen)$ (132)	Propylene	24 (rrrr)	[177]
	$Me_2Si(Cp)(C_5Me_4)$ (133)	Propylene	14 (rrrr)	[177]
	$Me_2C(Cp)(FluBu_2^i-2,7)$ (134)	Propylene	92 (rrrr)	[180]
	$MePhC(Cp)(Flu)$ (135)	Propylene	97 (rrrr)	[181]
	$(C_4H_8)C(Cp)(Flu)$ (136)	Propylene	92 (rrrr)	[182]
Hf	$Me_2C(Cp)(Flu)$ (130)	Propylene	74 (rrrr)	[64]
$V(acac)_3/AlClEt_2$ (137)		Propylene	65 (rr)	[178]
$V(mmh)_3/AlClEt_2$ (138)		Propylene	63 (rr)	[179]
Zr	$Me_2C(Cp)(Flu)$ (48)	1-Butene	92 (rr)	[176]
	Cp_2^* (139)	1-Butene	68 (rr)	[176]
Hf	Cp_2^* (140)	1-Butene	77 (rr)	[176]

4.2.3 Catalyst with or Without a Cp Ligand

Polymerization of propylene catalyzed by vanadium systems, e.g., $V(acac)_3/AlClEt_2$ (137) or $V(mmh)_3/AlClEt_2$ (138), mmh = 2-methyl-1,3-bu-tanedionato, at $-40°C$ slowly yielded a syndiotactic polymer, and this was achieved by chain-end control [178, 179]. ^{13}C NMR analysis of syndiotactic polypropylene obtained by using VCl_4-$Al(^{13}CH_3)_2Cl$ or VCl_4-$Al(^{13}CH_2CH_3)_3$ confirmed that the regiospecificity is a first-order Markov process [183].

Ishihara and co-workers discovered a new catalyst system, consisting of titanium halides or alkoxides with or without a cyclopentadienyl ligand in the presence of MAO, which provide highly active and stereoselective catalysts for highly syndiospecific polymerization of styrene [184, 185]. The catalytic activity depends on the substituents on the Cp ring, e.g. the tetramethylcyclopentadienyl ligand gives rise to the highest activity among $Ti(O^iPt)_3Cp$ complexes [186]. MAO-free systems derived from TiR_3Cp^* with $B(C_6F_5)_3$ were found to be catalysts for the syndiospecific polymerization of styrene as well as the poly-merization of ethylene and propylene [187, 188]. Recently two mechanisms have been proposed for this syndiospecific polymerization; (1) a carbocationic mecha-nism and (2) a cationic metal-center mechanism.

The first mechanism was proposed by Baird et al. [189]. The carbocationic species is schematically shown in Scheme IX. The attack of monomer, well known on the carbocationic center of a metal-ion-activated olefin, proceeded in the normal manner for carbocationic polymerization. This mechanism is based on the following two evidences. Alcoholysis of the polymerization system, $TiMe_3Cp^*/B(C_6F_5)_3$, resulted in the presence of an alkoxy group at an end group, and vinyl ethers and N-vinylcarbazole were polymerized by using the same system.

The latter mechanism is supported by evidence obtained from the initiation and termination steps in the syndiospecific polymerization of styrene [190]. The ^{13}C-enriched titanium catalyst afforded polystyrene with a $\cdots CH(Ph)CH_2{}^{13}CH_3$ end group, which indicates that the initiation step proceeded by secondary insertion (2,1-insertion) of styrene into the Ti-^{13}C bond of the active species (Eq. 10). In contrast to this mechanism, termination by the addition of ^{13}C-enriched methanol or tert-butyl alcohol afforded polymers without $^{13}CH_3O$ or *tertbutoxy* end groups.

$$Ti^{13}CH_3 \xrightarrow[\text{2,1-insertion}]{H_2C=CHPh} TiCH(Ph)CH_2{}^{13}CH_3$$

$$\xrightarrow[\text{2,1-insertion}]{n \; H_2C=CHPh} Ti[CH(Ph)CH_2]_n CH(Ph)CH_2{}^{13}CH_3 \qquad (10)$$

Ti = active site generated from $TiMe_3Cp^*$ and $B(C_6F_5)_3$

Several cationic titanium and zirconium monocyclopentadienyl derivatives, $[Cp^*MR_2]^+$ (M = Ti, Zr), have been synthesized by the reaction of MR_3Cp^*

$[CH_3-B(C_6F_5)_3]^-$ **Scheme 9.**

and $B(C_6F_5)_3$ [191, 192]. A benzyl derivative, $[Cp*Zr(CH_2Ph)_2][B(CH_2Ph)(C_6F_5)_3]$, is active for the polymerization of ethylene, and it oligomerizes propylene to give atactic oligomer [192]. A Cp-free zwitterionic zirconium complex, $Zr(CH_2Ph)_3\{(\eta^6\text{-}PhCH_2)B(C_6F_5)_3\}$, promotes polymerization of ethylene (25 kgPE(mol of $Zr)^{-1}h^{-1}atm^{-1}$) and propylene (1.5 kgPP(mol of $Zr)^{-1}h^{-1}atm^{-1}$) [193]. This complex stoichiometrically reacts with α-olefins, such as propylene, 4-methyl-1pentene, 1-vinylcyclohexane, and allylbenzene, to give the corresponding adducts, $[Zr(CH_2Ph)_3(CH_2CHRCH_2Ph)][B(C_6F_5)_3(CH_2Ph)]$, which polymerize propylene to give a mixture of atactic and isotactic polymers [194, 195].

Catalyst systems based on titanium compounds, e.g., CpTiCl$_3$, CpTiCl$_2$, Ti(CH$_2$Ph)$_4$, Ti(OR)$_4$, Ti(acac)$_3$, or TiPh$_2$, activated with MAO or B(C$_6$F$_5$)$_3$, were reported to give syndiotactic polystyrene [188, 196, 197]. During the catalysis, the cationic $[Cp*TiR_2]^+$ species are spontaneously reduced to Ti(III) species, which are now believed to be the true catalytic species for the syndiospecific polymerization of styrene [198, 199]. Some other organometallic systems, e.g. $\{N,N'\text{-Bis(trimethylsilyl)benzamidato}\}$titanium complexes/MAO such as $\{PhC(NSiMe_3)_2\}Ti(OPr^i)_3$, $[\{PhC(NSiMe_3)_2\}TiCl_3]_2$, $\{PhC(NSiMe_3)_2\}$TiCl$_3$L (L = thf, PMe$_3$) are also catalysts for the syndiospecific polymerization of styrene (93–95 wt% of s-PS, insoluble in refluxing 2-butanone) [200]. An example of a $\{N,N'\text{-dimethyl-}p\text{-toluamidato}\}$ derivative, $\{MeC_6H_4C(NMe)_2\}$TiCl$_3]_2$/MAO, has less stereospecificity [201].

The Dow corporation has recently developed constrained geometry addition polymerization catalysts (CGCT), typically $Me_2Si(C_5Me_4)(NBu^t)MCl_2$ (M = Ti, Zr, Hf) (**141**) activated with MAO. The homo-polymerization of α-olefins by CGCT afford atactic or somewhat syndiotactic (polypropylene: rr ≈ 69%) polymers. The metal center of the catalyst opens the coordination sphere and enables the co-polymerization of ethylene to take place, not only with common monomers such as propylene, butene, hexene, and octene, but also with sterically hindered α-olefins such as styrene and 4-vinylcyclohexene [202].

(M = Ti, Zr, Hf)

141

4.3 Theoretical Studies on the Stereoselectivity of Polymerization

Recently, theoretical calculations were done on the olefin polymerization. In particular, an ab-initio molecular orbital calculation was used to optimize the geometry for the ground, transition and product states of model systems, based on gas-phase reactions:

$$Cl_2MMe^+ \quad + \quad C_2H_4 \quad \longrightarrow \quad Cl_2MC_3H_7^+ \tag{11}$$

where $M = Ti$ and Zr [12, 203]. The model systems calculated by an energy-minimizing ab-initio method include bridging bis(cyclopentadienyl) moieties [13]. The most important optimized intermediate is the cationic ethylene π-complex, where the bond distances between Zr and the ethylene carbon atoms are 2.90 and 2.97 Å. These bond distances are longer than those (2.373(8) and 2.344(8) Å) of $Cp_2Zr(C_2H_4)(PMe_3)$ [27]. Thus, the ethylene coordination in the d^0 metallocene alkyl moiety is shown to be definitely weaker than that of Zr(II) complex.

The transition state was shown to have a four-centered nonplanar structure and the product showed a strong β-agostic interaction.[59] Molecular-mechanics (MM) calculations based on the structure of the transition state indicated that the regioselectivity is in good agreement with the steric energy of the transition state rather than the stability of the π-complex. The MM study also indicated that the substituents on the Cp rings determine the conformation of the polymer chain end, and the fixed polymer chain end conformation in turn determines the stereochemistry of olefin insertion at the transition state.[59]

The course of stereospecific olefin polymerization was studied by using the molecular mechanics programs, MM-2 and Biograph, based on the optimized geometries of the ethylene complex and the transition state [13, 203]. Interestingly, the steric interaction at the transition state mainly controls the stereochemistry in polymerization, which proceeds specifically isotactic or syndiotactic depending on the kind of catalyst.

5 Polymerization of Non-Conjugated Dienes by Organometallic Complexes

Recently, a metallocene/MAO system has been used for the polymerization of non-conjugated dienes [204, 205]. The cyclopolymerization of 1,5-hexadiene has been catalyzed by Zieger-Natta catalyst systems, but with low activity and incomplete cyclization in the formation 5-membered rings [206]. The cyclopolymerization of 1,5-hexadiene in the presence of $ZrMe_2Cp_2/MAO$ afforded a polymer ($M_w = 2.7 \times 10^7$, $M_w/M_n = 2.2$) whose NMR indicated that almost complete cyclization had taken place. One of the olefin units of 1,5-hexadiene is initially inserted into an M–C bond and then cyclization proceeds by further

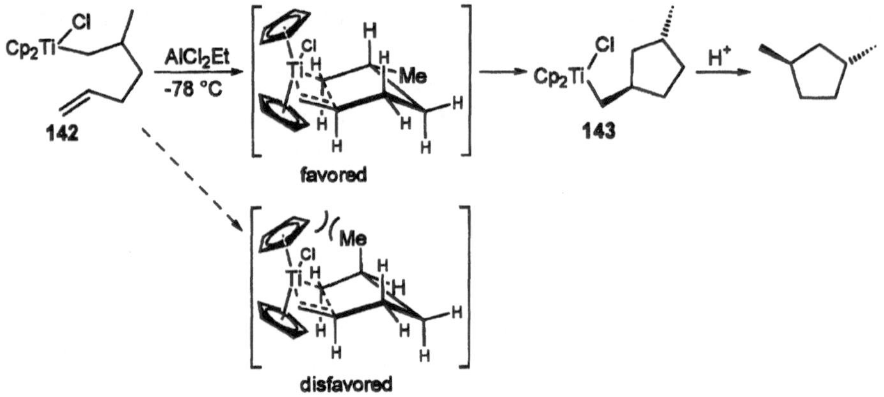

Scheme 10.

insertion of the remaining unit. The stereochemistry of the polymer containing 1,3-disubstituted cyclopentane units was made trans-selective via a pseudo-chair transition state. The selectivity was improved by decreasing the reaction temperature.

In some of the model reactions, the same stereochemistry was confirmed. Thus, intramolecular olefin insertion occurred with exo-regioselectivity. For example, treatment of $Cp_2Ti(Cl)(CH_2CHMeCH_2CH_2CH = CH_2)$ (**142**) with EtAlCl$_2$ produced **143**, which resulted in the formation of a 92:8 ratio of *trans:cis* 1,3-dimethylcyclopentane upon protonolysis (Scheme X) [207]. Without the Cp ligand, the polymerization of 1,5-hexadiene by using TiCl$_3$/EtAlCl$_2$ as the catalyst system yielded a polymer with a 46:54 ratio of *trans:cis* disubstituted cyclopentane rings [208]. More sterically hindered $Cp_2^*ZrCl_2$ was used as a catalyst component to give a high-*cis* polymer[80]. The origin of this stereoselectivity is not clear, but twist-boat conformation at the transition state could be the explanation.

When a chiral *ansa*-type zirconocene/MAO system was used as the catalyst precursor for polymerization of 1,5-hexadiene, an main-chain optically active polymer (68% *trans* rings) was obtained[84–86]. The enantioselectivity for this cyclopolymerization can be explained by the fact that the same prochiral face of the olefins was selected by the chiral zirconium center (Eq. 12) [209–211]. Asymmetric hydrogenation, as well as C–C bond formation catalyzed by chiral *ansa*-metallocene **144**, has recently been developed to achieve high enantioselectivity[88–90]. This parallels to the high stereoselectivity in the polymerization.

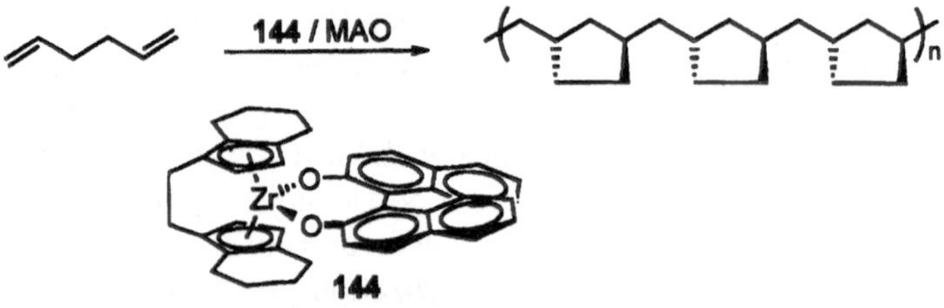

$$(12)$$

6 Polymerization of Functionalized Olefins

Ziegler-Natta catalyst systems containing an excess of alkylaluminum have been found to be unstable catalysts for the polymerization of the polar monomers and functionalized olefins. For example, a $TiCl_4$-Et_3Al catalyst system does not polymerize MMA above 0°C but polymerizes it only at low temperatures, such as -78°C, to give syndiotactic PMMA [212, 213]. In contrast, cationic Group 4 and neutral Group 3 metallocene complexes can be applied for this purpose above 0°C. A catalyst derived from $Cp_2^*ZrMe_2$ and $B(C_6F_5)_3$ was reported to be active for the polymerization of functionalized α-olefins containing silyl-protected alcohols or tertiary amines [214].

The highly syndiospecific-living polymerization of methyl methacrylate has been initiated by the neutral bis(pentamethylcyclopentadienyl)lanthanide-alkyl or -hydride complexes [215, 216]. The plausible reaction mechanism is shown in Scheme XI.

At the first step, the insertion of MMA to the lanthanide–alkyl bond gave the enolate complex. The Michael addition of MMA to the enolate complex via the 8-membered transition state results in stereoselective C–C bond formation, giving a new chelating enolate complex with two MMA units; one of them is enolate and the other is coordinated to Sm via its carbonyl group. The successive insertion of MMA afforded a syndiotactic polymer. The activity of the polymerization increased with an increase in the ionic radius of the metal (Sm > Y > Yb > Lu). Furthermore, these complexes become precursors for the block co-polymerization of ethylene with polar monomers such as MMA and lactones [215, 217].

The crystal structure of the 1:2 adduct **145** obtained by the reaction of $[Cp_2^*SmH]_2$ with MMA was determined by X-ray analysis. One of the two monomer units is in the *O*-enolate form and the other unit coordinates to the samarium atom by the carbonyl group. A comparison between **145** and the

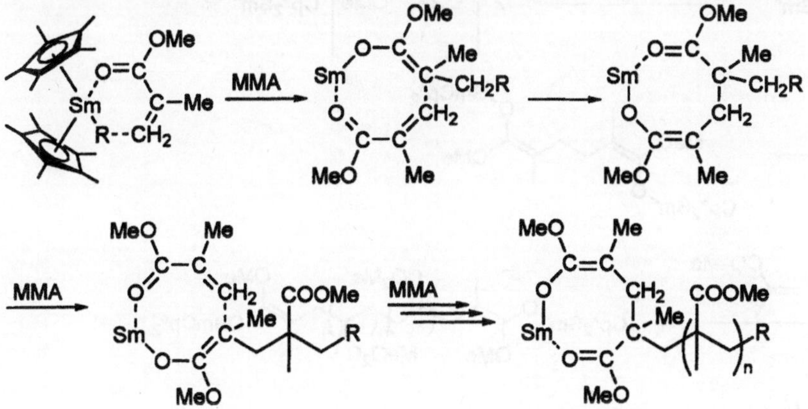

Scheme 11.

analogous rhodium complex **146** is worth considering since they are intermediates in the catalytic tail-to-tail dimerization of MA [218, 219]. The oxophilic properties of the early transition metals leads to the enolate complex, whereas the rhodium complex catalyzes the insertion of MA into an Rh-alkyl complex and β-hydrogen elimination regenerates active hydride species and the dimer of MA [220].

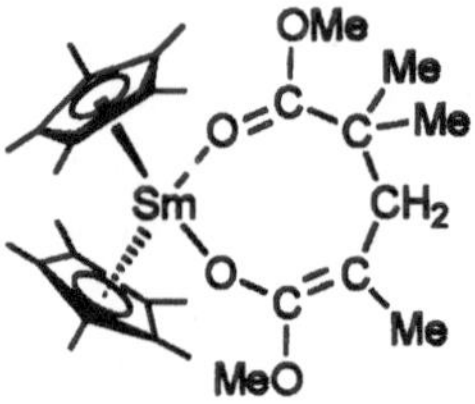

145 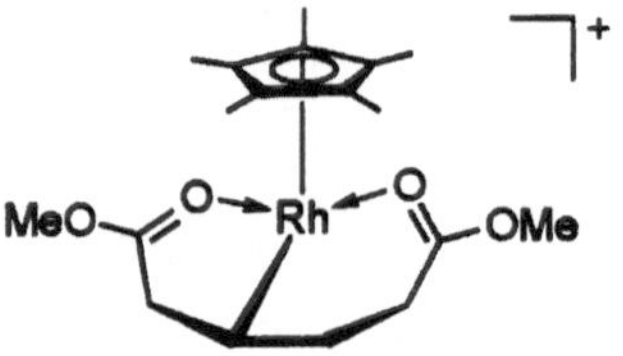　　　　　　　**146**

Divalent samarium and ytterbium complexes, $Cp^*_2Ln(thf)_2$ (Ln = Sm, Yb), also initiate the syndiospecific living polymerization of MMA although the initiator efficiencies are rather low [216]. Novak et al. proposed an initiation mechanism that includes the generation of the MMA radical anion, followed by coupling to form a bifunctional initiator for this catalyst system (Scheme XII) [221]. A bimetallic diyne complex, $C^*_2Sm(PhC_4Ph)C^*_2Sm$, was found to be an efficient bisinitiator. A sterically less-demanding ytterbocene derivative, $(MeC_5H_4)_2 Yb(dme)$, was found to have higher activities than the Cp* derivatives [222].

Similar polymerization of MMA using enolate-zirconocene catalysts has also been found [223]. The mechanism of this catalytic reaction is related to the process described in Scheme XI because the cationic enolate complex is isolobal to that of the corresponding lanthanide complex. Recently, similar cationic

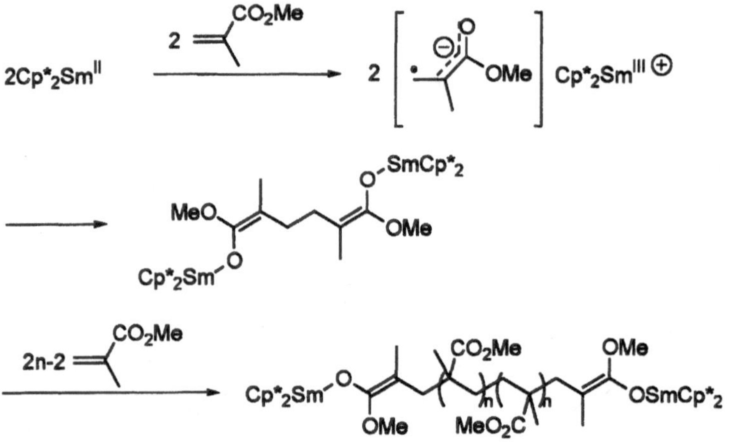

Scheme 12.

zirconocene species formed by the combination of Cp_2ZrX_2 with $[Ph_3C][B(C_6F_5)_4]$ and diethylzinc was found to catalyze the living polymerization of MMA. In addition, a racemic *ansa*-type zirconocene precursor in the same condition has been reported to catalyze the *isotactic* polymerization of MMA [224, 225]. Chiral *ansa*-type complexes of rare earth elements bearing a chiral auxiliary ligans can catalyze the isospecific polymerization of MMA [226]. Some non-metallocene complexes of lanthanide metals such as SmI_2 [227] and $Ln(SAr)_nL_x$ (Ln = Sm, Eu, Yb; SAr = SPh, $SC_6H_3Pr^i_3$-2,4,6; L = HMPA, THF, pyridine) [228] were found to catalyze the polymerization of MMA. Both iodide and thiolate ligands are less sterically demanding and less electron-donating than Cp*, the active species seems to be somewhat unstable. We found that the catalyst system $Yb(SPh)_3(hmpa)_3/3AlMe(dbmp)_2$ (dbmp = 2,6-di-*tert*-butyl-4-methylphenoxide) affords highly syndiotactic PMMA with a relatively narrow molecular weight distribution (in THF at $-78°C$, $M_n = 1.01 \times 10^5$, $M_w/M_n = 1.17$, rr = 87%) [228].

7 Recent Trends in Polymerization Catalyzed by Organometallic Complexes of Transition Metals Other than Those of Group 4

7.1 Polymerization by Organometallic Complexes of Group 5 Metals

Metallocene complexes, $NbCl_3Cp_2$, $NbCl_2Cp_2$, and half metallocene complexes, $NbCl_4Cp$, showed no catalyst activity even in the presence of MAO [229]. Neutral metallocene alkyl complexes of Group 3 and cationic metallocene alkyl complexes of Group 4 metals are 14 electron species and considered to be in isolobal and isoelectronic relationships (Chart 2). From this viewpoint, we suggested that 14-electron cationic species of Group 5 metals, i.e. $[MRCp(1,3-diene)]^+$ (M = Nb and Ta), should be active catalysts for the polymerization of alkenes. Thus, the combination of $MX_2Cp(\eta^4$-1,3-diene) and MAO was examined to find the living polymerization of ethylene [230–233].

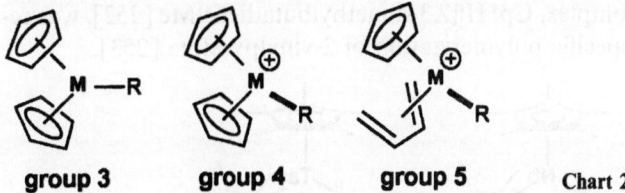

group 3   group 4   group 5   Chart 2

Similar isolobal and isoelectronic relationships [234, 235] have recently been observed for various organometallic complexes of both Group 5 metals

[236–239] and Group 6 metals [240–242]. Actually we prepared ethylene [243], benzyne [244], and benzylidene complexes [245] bearing a TaCp* (η^4-buta-1,3-diene) fragment; they may be compared with the metallocene complexes of Group 4 metals. Thus, we found that the system of $MX_2(\eta^5\text{-}C_5R_5)(\eta^4\text{-diene})$ (M = Nb and Ta; R = H and CH_3; X = Cl and CH_3) in the presence of an excess of MAO is a catalyst precursor for the living polymerization of ethylene; and the narrowest polydispersity (M_w/M_n as low as 1.05) for polyethylene was accomplished at lower temperature (Table 10) [246–248]. The catalytic activity delicately depends on the steric and combined electronic effects of the cyclopentadienyl and diene ligands. Thus, the optimum temperature for the polymerization reaction varies, depending on the nature of those ligands.

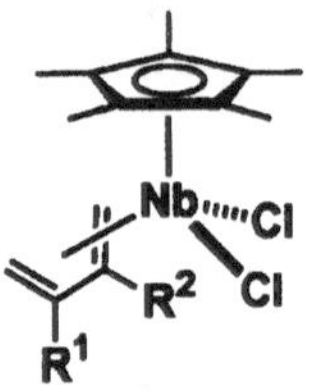

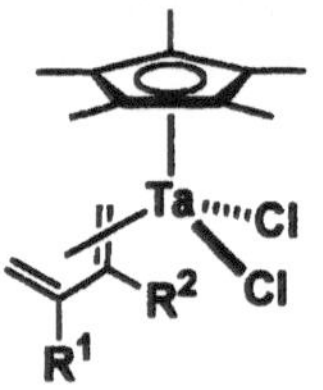

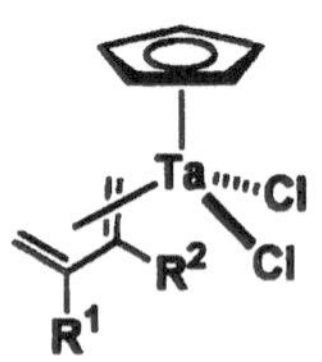

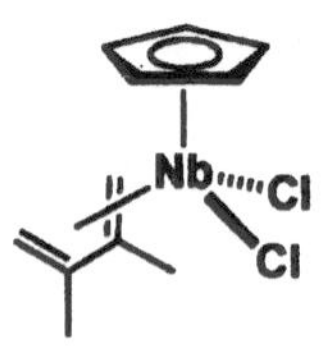

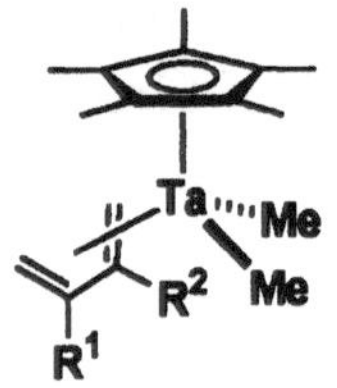

a: $R^1 = R^2 = H$
b: $R^1 = CH_3$, $R^2 = H$
c: $R^1 = R^2 = CH_3$

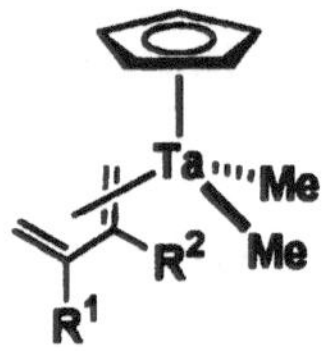

We found that bis(diene) complexes of niobium and tantalum, $MCp^*(\eta^4\text{-diene})_2$, were also active catalyst precursors for the polymerization of ethylene when MAO was used as cocatalyst (Table 11) [247, 248]. This might be attributed to the stability of the diene aluminum species [249] generated by the ligand exchange reaction, and the generated cationic species should be the same as that derived from $MX_2Cp(\eta^4\text{-diene})$ and MAO. The second diene ligand is coordinated to the metal that has a higher σ-bond character like that of the zirconocene diene complexes. Similarly, a combination of diene complexess of zirconocene with MAO [250] or $B(C_6F_5)_3$ [251] was found to be the catalyst precursor for the polymerization of α-olefins. A mono(cyclopentadienyl) mono(diene)hafnium complex, Cp*Hf(2,3-dimethylbutadiene)Me [252], was reported to catalyze isospecific polymerization of 2-vinylpyridine [253].

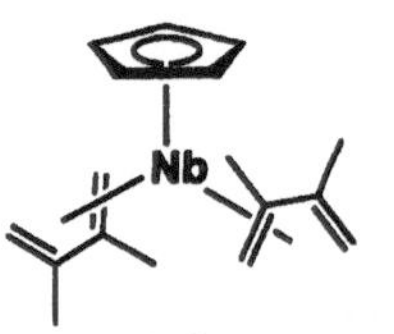

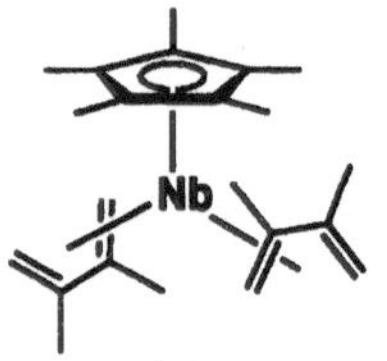

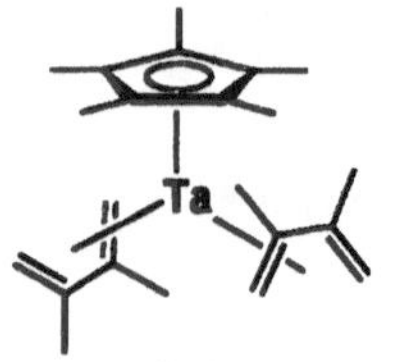

Table 10. Polymerization of ethylene catalyzed by niobium-diene and tantalum-diene complexes/MAO[a]

Complex	Time (h)	Temperature (°C)	Activity[b] (kg/h·[M]mol)	$M_n/10^4$ [c]	M_w/M_n [c]
147a	1	20	38.70	8.29	1.30
147a	1	− 20	10.65	2.36	1.05
147b	1	20	19.22	3.95	1.16
147b	1	− 20	1.02	0.51	1.09
147c	1	20	35.23	10.54	1.18
147c	1	− 20	12.71	4.10	1.05
148a	6	20	5.93	2.03	2.04
148a	6	− 20	1.51	2.03	1.16
148b	6	20	1.90	1.15	1.63
149a	6	20	0.52	–	–
149a	6	− 20	7.07	8.18	1.40
149b	6	20	0.88	–	–
149c	6	20	3.77	–	–
150	3	20	6.89	–	–
151a	6	20	4.48	2.04	2.06
151a	6	− 20	1.18	2.55	1.08
151b	6	20	1.61	1.23	1.65
151c	6	20	5.69	1.20	2.09
152a	6	20	0.37	–	–
152a	6	− 20	5.47	8.74	1.51

[a] Polymerization reactions were carried out in toluene (1.44×10^{-3} M of [M]) in the presence of MAO (500 equiv).
[b] Polymer was methanol-insoluble parts.
[c] GPC analysis.

Table 11. Polymerization of ethylene catalyzed by bis-diene complexes/MAO[a]

Complex	Time (h)	Temperature (°C)	Activity[b] (kg/h·[M]mol)	$M_n/10^4$ [c]	M_w/M_n [c]
153	1	20	9.03	16.86	2.24
153	1	0	17.03	31.80	2.94
153	1	− 20	7.55	27.14	4.13
154	1	20	24.65	3.07	1.18
154	1	0	21.95	2.82	1.13
154	1	− 20	12.48	1.42	1.06
155	6	20	3.39	0.40	2.00
155	6	0	1.44	0.96	1.30
155	6	− 20	0.26	0.54	1.09

[a] Polymerization reactions were carried out in toluene (1.44×10^{-3} M of [metal]) in the presence of MAO (500 equiv).
[b] Polymer was methanol-insoluble parts.
[c] GPC analysis.

Two other isoelectronic tantalum complexes, **156** [254] and **157** [255], have recently been reported to show a similar catalytic activity for the polymerization of ethylene (2 kg PE/h[Ta]mol and 12 kg PE/h[Ta]mol, respectively) when they were activated with MAO.

156   **157**

The isoelectronic relationship has been applied to the vanadium complex with hydrotris(pyrazolyl)borato ligand, which possesses the same electronic features as the Cp ligand; thus the combination of complex $VCl_2(NAr)Tp^*$ **(158)** (Ar = 2,6-diisopropylphenyl, Tp^* = hydrotris(3,5-dimethylpyrazolyl) borate) with MAO was reported to catalyze the polymerization of ethylene (at atmospheric pressure) and propylene (at 7 bar), giving polyethylene (14 kg/mol·h, M_w = 47000, M_w/M_n = 3.0) and polypropylene (1 kg/mol·h, M_w = 3800, M_w/M_n = 2.0), respectively [256].

158

The vanadium complex **159**, bearing oligometallasilsesquioxanes, is an interesting model complex for surface species on silica [257]. Polymerization of ethylene with **159**/MAO affords polyethylene with an M_n value of 2100 and an M_w value of 47900. An aluminum-vanadium adduct **160** polymerized ethylene without any cocatalysts, affording similar catalytic activity with the **159**/MAO system [258].

159   **160**

7.2 Polymerization by Organometallic Compounds of Group 6 Metals

Although chromium-based ethylene polymerization catalysts have already been developed commercially, these processes are based on a heterogeneous catalyst system [259, 260]. Recently, it was found that soluble organochromium complexes are able to catalyze ethylene polymerization [261–263]. A cationic chromium(III) complex $[Cp*CrMe(THF)_2]^+(BPh_4)^-$ (161) in dichloromethane catalyzed ethylene polymerization, although 161 is not active for propylene polymerization. A kinetic study showed that the pseudo-five-coordinated 13e $[Cp*CrMe(THF)]^+$ (162) is the active species. Thus, the addition of THF to 161 inhibited the polymerization, suggesting that the coordination of ethylene to the alkylchromium complex is the key intermediate to successive ethylene insertions into the Cr–C bond. The oxidation state of chromium seems to be important for achieving high activity. Two alkylchromium compounds, $[Cp*Cr(dmpe)Me]^+$ (163) and $Cp*Cr(dmpe)Me$ (164), are isostructural but differ in their formal oxidation state, Cr(III) for 163 vs. Cr(II) for 164. Complex 163 is a catalyst precursor for ethylene polymerization, whereas the reaction of 164 with ethylene afforded mainly propylene, which was formed by the insertion of ethylene into the Cr(II)–Me bond followed by β-hydrogen elimination. Such a large difference between 163 and 164 may be attributed to the fact that olefins can coordinate strongly to a more electron-rich Cr(II) d^4 center via enhanced π-back bonding, as described in the previous section for organometallic tantalum complexes.

Some benzyl complexes of chromium – $Cp*Cr(CH_2Ph)_2(thf)$, $[Cp*Cr(CH_2Ph)(thf)_2][BPh_4]$ and $Li[Cp*Cr(CH_2Ph)_3]$ – catalyzed the polymerization of ethylene $[0.11–0.28\,kgPE(mol$ of $Cr)^{-1}h^{-1}atm^{-1}]$ [264]. Neutral alkylchromium(III) complexes such as $Cr(CH_2SiMe_3)_2Cp*$ were also found to catalyze the rapid polymerization of ethylene at $-40°C$. Thus, both neutral and cationic coordinatively unsaturated alkyl complexes of chromium(III) are the active catalyst for ethylene polymerization. Quite recently, the well-defined 16-e complexes of chromium $(RN)_2CrX_2$ $(R = Bu^t$, Ph; $X = CH_2Ph$, Cl), which are the isoelectronic analogues of Cp_2TiX_2, activated by $[Ph_3C][B(C_6F_5)_4]$, have been reported to catalyze the polymerization of ethylene [265].

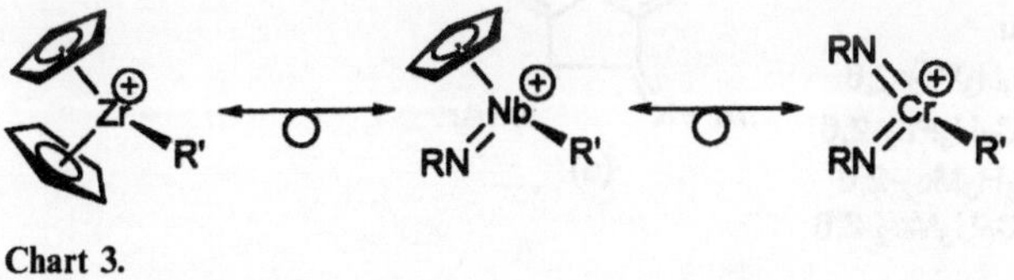

Chart 3.

In the case of other Group 6 metals, the polymerization of olefins has attracted little attention. Some molybdenum(VI) and tungsten(VI) complexes containing bulky imido- and alkoxo-ligands have been mainly used for metathesis reactions and the ring-opening metathesis polymerization (ROMP) of norbornene or related olefins [266–268]. Tris(butadiene) complexes of molybdenum(0) and tungsten(0) are air-stable and sublimable above 100°C [269, 270]. At elevated temperature, they showed catalytic activity for the polymerization of ethylene [271].

7.3 Polymerization Assisted by Organometallic Compounds of Other Transition Metals

Organometallic compounds of Group 8–10 metals have been considered as the catalysts for olefin dimerization and oligomerization [272]. These metals prefer β-hydrogen elimination followed by reductive elimination. One exception reported so far is the SHOP system, in which nickel catalysts produce higher olefins [273]. Brookhart et al. have reported a living polymerization of ethylene that is accomplished by a combination of a cobalt olefin complex, $(C_5Me_5)Co\{P(MeO)_3\}$ $(H_2C = CHAr)$ $(Ar = Ph, p-C_6H_4Cl, p-C_6H_4OAc, p-C_6H_4CF_3, C_6F_5)$, with $HB\{C_6H_3(CF_3)_2-3,5\}_4$ [274]. In this catalyst system, a mechanism involving β-agostic interaction as a resting state of propagation has been proposed, though the activities were rather low $(Ar = C_6H_4CF_3$, activity 3.8 kg(mol of catalyst)$^{-1}h^{-1}atm^{-1})$, $M_n = 1.48 \times 10^4$, $M_w/M_n = 1.11)$ [274]. Another catalyst system based on nickel or palladium with 1,4-diazadiene derivatives (165, 166) was found to polymerize α-olefins, when the unique combination of a chelating nitrogen ligand and coordinatively weak anion $[B(C_6H_3-3,5-(CF_3)_2)_4]^-$ was carefully selected. Thus, polymerization of ethylene by palladium catalysts 165 afforded extremely branched polyethylene (> 100 branches per 1000 carbon atoms) with high molecular weight. A nickel catalyst system exhibited extremely high activity for the polymerization of ethylene (activity of 166a: 3600 kg(mol of catalyst)$^{-1}h^{-1}atm^{-1}$) [275]. When complex 167a was activated by MAO, higher activity (activity of 167a/MAO: 11000 kg(mol of catalyst)$^{-1}h^{-1}atm^{-1}$) was observed [275].

165 M = Ni
166 M = Pd

167 M = Ni

(a) R = H, Ar = $C_6H_3Pr^i_2$-2,6
(b) R = Me, Ar = $C_6H_3Pr^i_2$-2,6
(c) R = H, Ar = $C_6H_3Me_2$-2,6
(d) R = Me, Ar = $C_6H_3Me_2$-2,6

(e)

The copolymerization of carbon monoxide and α-olefins is one of the most challenging problems in polymer synthesis. Sen and his coworkers discovered that some cationic palladium compounds catalyze this alternative copolymerization, giving polyketones (Eq. 13).

$$R \diagup\diagdown \;+\; CO \xrightarrow{\text{Chiral Pd}^{II}\text{ Catalyst}} \qquad\qquad \tag{13}$$

This polymer is now attracting much attention because of its potential as a photodegradable polymer [276, 277]. A mechanism involving the carbonylation of cationic alkylpalladium on the basis of the isolated acyl complex has been proposed [278]. It is noteworthy that the cationic alkyl palladium complex with chelating nitrogen ligands exhibited excellent activity. Thus, some research groups of Sen [279–281], Brookhart [282], and Consiglio [283] have used chiral chelating phosphine or oxazoline ligands to introduce main-chain chirality. An optically active alternating copolymer of α-olefin and carbon monoxide was obtained by using chiral palladium(II) complexes as catalysts **168–170**. In the catalyst system of **168**, the catalysts bearing planar achiral ligands, **168a**, **168b**, and **168c**, gave syndiotactic poly(styrene-*alt*-CO) by chain-end control [282]. The use of C_2-symmetric chiral ligands in **168e** provides enantiomorphic site control to produce highly isotactic, optically active copolymers [282]. The novel alternating isomerization/co-oligomerization of 2-butene with carbon monoxide catalyzed by **169d** was found to give optically active, isotactic poly(1,5-ketone) (Eq. 14) [281].

$$\sim\!\!\!\!/ \quad + \quad CO \quad \xrightarrow{\quad 169d \quad} \quad \left(\!\!\! \begin{array}{c} \\ \end{array}\!\!\!\right)_n \qquad (14)$$

A unique chiral phosphine-phosphate ligand has been applied for hydroformylation reaction of alkenes. In the case of **170** this ligand can be applied for obtaining poly(propylene-*alt*-Co) with the highest molar optical rotation and with the highest molecular weight ($[\Phi]_D^{24} = +40°$, $M_w = 104400$, $M_w/M_n = 1.6$) [284].

8 Concluding Remarks

The features of homogeneous metallocene/MAO catalysts distinguishable from the classical heterogeneous Ziegler-Natta catalysts can be summarized as follows:
1) They are single-site catalysts.
2) Narrow molecular weight distribution of the resulting polymer can be obtained.
3) Highly random co-polymerization of α-olefins is possible.
4) They are active not only to small olefins but also to higher or bulky olefins.
5) The stereoregularity and the molecular weight distribution of the resulting polymer can be controlled by the design of the catalyst precursor.
6) Some polar vinyl monomers such as MMA can be polymerized.

The initial Group 4 metallocene/MAO catalyst systems has some disadvantages:
7) MAO co-catalyst is expensive but required in large excess amount.
8) The molecular weights of the resulting polymers tend to be lower.
9) It is difficult to obtain the polymer as particles with uniform size.

At present, the difficulties described in 7–9 can be solved. The use of MAO can be avoided by the use of non-MAO catalysts such as cationic Group 4 metallocenes or neutral Group 3 metallocenes. It has been found that the molecular weight of the resulting polymer can be increased by introducing substituents at the 2,5-position in the cyclopentadienyl group of the *ansa*-metallocenes. When the metallocene catalysts are heterogenized by being supported on appropriate materials, they can be used in slag or gas-phase polymerization systems to yield polymers in the form of uniform particles with stereoregularity corresponding to that of homogeneous systems [285–287].

The molecular design of stereospecific homogeneous catalysts for polymerization and oligomerization has now reached a practical stage, which is the result of the rapid developments in early transition metal organometallic chemistry in this decade. In fact, Exxon and Dow are already producing polyethylene commercially with the help of metallocene catalysts. Compared to the polymerization of α-olefins, the polymerization of polar vinyl, alkynyl and cyclic monomers seems to be less developed.

Non-metallocene complexes, such as aryloxide **31** and amide **138**, have also been utilized as catalyst systems for the polymerization of α-olefins. Moreover, the homogeneous olefin polymerization catalysts have been extended to metals other than those in Group 4, as described in Sect. 7. Complexes such as mono(cyclopentadienyl)mono(diene) are in isoelectronic relationship with Group 4 metallocenes and they have been found to initiate the living polymerization of ethylene. These studies will being further progress to the chemistry of homogeneous polymerization catalysts.

The worldwide research activities in this field of organometallic catalysts for polymerization have been maintained at the highest level for years. Many more examples will appear in the near future and the whole scope of organometallic olefin polymerization chemistry will be clarified.

Acknowledgements. We are grateful for financial support from the Ministry of Education, Science, Sports, and Culture of Japan (Grant-in-Aid for Scientific Research on Priority Areas of Reactive Organometallics No. 05236106 and Specially Promoted Research No. 06101004). K. M. appreciates the financial support from the Kurate Foundation.

9 References

1. Fink, G, Mülhaupt R, Brintzinger HH (1995) Ziegler Catalysts. Springer, Berlin Heidelberg New York
2. Keii T, Soga K (1990) Catalytic Olefin Polymerization. Elsevier, Lausanne
3. Quirk RP (1988) Transition-Metal Catalyzed Polymerizations; Ziegler-Natta and Metathesis Polymerizations. Cambridge University Press, Cambridge
4. Kaminsky W, Sinn H (1988) Organometallics as Catalysts for Olefin Polymerization. Springer, Berlin Heidelberg New York
5. Brintzinger HH, Fischer D, Mülhaupt R, Rieger B, Waymouth RM (1995) Angew Chem, Int Ed Engl 34: 1143
6. Grubbs RH, Coates GW (1996) Acc Chem Res 29: 85
7. Bochmann M (1996) J Chem Soc, Dalton Trans 255
8. Sinn H, Kaminsky W (1980) Adv Organomet Chem 18: 99
9. Zakharov VA, Bukatov GD, Yermakov YI (1983) Adv Polym Sci 51: 61
10. Kaminsky W, Steiger R (1988) Polyhedron 7: 2375
11. Casey CP, Hallenbeck SL, Pollock DW, Landis CR (1995) J Am Chem Soc 117: 9770
12. Kawamura-Kuribayashi H, Koga N, Morokuma K (1992) J Am Chem Soc 114: 2359

13. Kawamura-Kuribayashi H, Koga N, Morokuma K (1992) J Am Chem Soc 114: 8687
14. Eichner ME, Alt HG, Rausch MD (1984) J Organomet Chem 264: 309
15. Steigerwald ML, Goddard III WA (1985) J Am Chem Soc 107: 5027
16. Kaminsky W, Kopf J, Sinn H, Vollmer H-J (1976) Angew Chem, Int Ed Engl 15: 629
17. Takahashi T, Kasai K, Suzuki N, Nakajima K, Negishi E-I (1994) Organometallics 13: 3413
18. Cotton FA, Kibala PA (1990) Inorg Chem 29: 3192
19. Bruns CJ, Andersen RA (1987) J Am Chem Soc 109: 915
20. Evans WJ, Ulibarri TA, Ziller JW (1990) J Am Chem Soc 112: 219
21. Burns CJ, Andersen RA (1987) J Am Chem Soc 109: 941
22. Bartell LS, Roth EA, Hollowell CD, Kuchitsu K, Young JE Jr (1965) J Chem Phys 42: 2683
23 Levy GC, Nelson GL (1972) Carbon-13 Nuclear Magnetic Resonance for Organic Chemists, Wiley-Interscience, New York
24. Lynden-Bell RM, Sheppard N (1962) Proc Roy Soc, Ser A 269: 385
25. Cohene SA, Auburn PR, Bercaw JE (1983) J Am Chem Soc 105: 1136
26. Takahashi T, Murakami M, Kunishige M, Saburi M, Uchida Y, Kozawa K, Uchida T, Swanson DR, Negishi E (1989) Chem Lett 761
27. Alt HG, Denner CE, Thewalt U, Rausch MD (1988) J Organomet Chem 356: C83
28. Goddard R, Binger P, Hall SR, Müller P (1990) Acta Crystallogra, Sect C 46: 998
29. Buchwald SL, Kreutzer KA, Fisher RA (1990) J Am Chem Soc 112: 4600
30. Spencer MD, Morse PM, Wilson SR, Girolami GS (1993) J Am Chem Soc 115: 2057
31. Guggenberger LJ, Meakin P, Tebbe FN (1974) J Am Chem Soc 96: 5420
32. Burger BJ, Santarsiero BD, Trimmer MS, Bercaw JE (1988) J Am Chem Soc 110: 3134
33. Arnold J, Don Tilley T, Rheingold A, Geib SJ (1987) Organometallics 6: 473
34. Poole AW, Gibson VC, Clegg W (1992) J Chem Soc, Chem Commun 237
35. Schultz AJ, Brown RK, Williams JM, Schrock RR (1981) J Am Chem Soc 103: 169
36. Abbenhuis HCL, Feiken N, Grove DM, Jastrzebski JTBH, Kooijman H, van der Sluis P, Smeets WJJ, Spek AL, van Koten G (1992) J Am Chem Soc 114: 9773
37. Wielstra Y, Meetsma A, Gambarotta S (1989) Organometallics 8: 258
38. Wielstra Y, Gambarotta S, Spek AL (1990) Organometallics 9: 572
39. Wreford (1980) J Organomet Chem 188: 353
40. Breslow DS, Newburg NR (1957) J Am Chem Soc 79: 5072
41. Sinn H, Kaminsky W, Jollmer HJ, Woldt R (1980) Angew Chem, Int Ed Engl 19: 390
42. Kaminsky W, Miri M, Sinn H, Woldt R (1983) Makromol Chem, Rapid Commun 4: 417
43. Herwig J, Kaminsky W (1983) Polym Bull 9: 464
44. Giannetti E, Nicoletti GM, Mazzocchi R (1985) J Polym Sci, Polym Chem Ed 23: 2117
45. Ewen JA, Haspeslagh L, Atwood JL, Zhang H (1987) J Am Chem Soc 109: 6544
46. Miya S, Mise T, Yamazaki H (1989) Chem Lett 1853
47. Busico V, Cipullo R (1994) J Am Chem Soc 116: 9329
48. Resconi L, Fait A, Riemontesi F, Colonnesi M, Rychlicki H, Zeigler R (1995) Macromolecules 28: 6667
49. Lee I-M, Gautheir WJ, Ball JM, Iyengar B, Collins S (1992) Organometallics 11: 2115
50. Spaleck W, Antberg M, Rohrmann J, Winter A, Bachmann B, Kiprof P, Behm J, Herrmann WA (1992) Angew Chem, Int Ed Engl 31: 1347
51. Linden Avd, Schaverien CJ, Meijboom N, Ganter C, Orpen AG (1995) J Am Chem Soc 117: 3008
52. Miyatake T, Mizunuma K, Seki Y, Kakugo M (1989) Macromol Chem, Rapid Commun 10: 349
53. Herrmann WA, Rohrmann J (1989) Angew Chem, Int Ed Engl 28: 1511
54. Hlatky GG, Eckman RR, Turner HW (1992) Organometallics 11: 1413
55. Crowther DJ, Baenziger NC, Jordan RF (1991) J Am Chem Soc 113: 1455
56. Jeske G, Lauke H, Mauermann H, Swepston PN, Schumann H, Marks TJ (1985) J Am Chem Soc 107: 8091
57. Watson PL, Parshall GW (1985) Acc Chem Res 18: 51
58. Herskovics-Korine D, Eisen MS (1995) J Organomet Chem 503: 307
59. Kaminsky W, Kulper K, Brintzinger HH, Wild FRWP (1985) Angew Chem, Int Ed Engl 24: 507
60. Mise T, Miya S, Yamazaki H (1989) Chem Lett 1853
61. Röll W, Brinzinger H-H, Rieger B, Zolk R (1990) Angew Chem, Int Ed Engl 29: 279

62. Collins S, Gauthier WJ, Holden DA, Kuntz BA, Taylor NJ, Ward DG (1991) Organometallics 10: 2061
63. Stehling U, Diebold J, Röll W, Brintzinger HH (1994) Organometallics 13: 964
64. Ewen JA, Jones RL, Razavi A, Ferrara JD (1988) J Am Chem Soc 110: 6255
65. Ewen JA (1984) J Am Chem Soc 106: 6355
66. Mallin DT, Rausch MD, Lin Y-G, Dong S, Chien JCW (1990) J Am Chem Soc 112: 2030
67. Chien JCW, Llinas GH, Rausch MD, Lin G-Y, Winter HH, Atwood JL, Bott SG (1991) J Am Chem Soc 113: 8569
68. Harlan CJ, Bott SG, Barron AR (1995) J Am Chem Soc 117: 6465
69. Gassman PG, Callstrom MR (1987) J Am Chem Soc 109: 7875
70. Sishta C, Huthorn RM, Marks TJ (1992) J Am Chem Soc 114: 1112
71. Toscano PJ, Marks TJ (1985) J Am Chem Soc 107: 653
72. Finch WC, Gillespie RD, Hedden D, Marks TJ (1990) J Am Chem Soc 112: 6221
73. Dahmen K-H, Hedden D, Burwell RLJ, Marks TJ (1988) Langmuir 4: 1212
74. Toscano PJ, Marks TJ (1986) Langmuir 2: 820
75. Jordan RJ, Bradely PK, LaPointe RE, Taulor DF (1990) New J Chem 14: 505
76. Jordan RF (1991) Adv Organomet Chem 32: 325
77. Eisch JJ, Piotrowski AM, Brownstein SK, Gabe EJ, Lee FL (1985) J Am Chem Soc 107: 7219
78. Bochmann M, Jaggar AJ, Wilson LM, Hursthouse MB, Motevalli M (1989) Polydedron 8: 1838
79. Yang X, Stern CL, Marks TJ (1991) J Am Chem Soc 113: 3623
80. Bochmann M, Lancaster SJ, Hursthouse MB, Malik KMA (1994) Organometallics 13: 2235
81. Jordan RF, Bajgur CS, Willett R, Scott B (1986) J Am Chem Soc 108: 7410
82. Amorose DM, Lee RA, Petersen JL (1991) Organometallics 10: 2191
83. Eshuis JJW, Tan YY, Meetsma A, Teuben JH, Renkema J, Evens GG (1992) Organometallics 11: 362
84. Razavi A, Thewalt U (1993) J Organomet Chem 445: 111
85. Jordan RF, LaPointe RE, Bajgur CS, Echols SF, Willett R (1987) J Am Chem Soc 109: 4111
86. Jordan RF, Bradley PK, Baenziger NC, LaPointe RE (1990) J Am Chem Soc 112: 1289
87. Alelyunas YW, Baenziger NC, Bradley PK, Jordan RF (1994) Organometallics 13: 148
88. Horton AD, Orpen AG (1992) Organometallics 11: 1193
89. Horton AD, Orpen AG (1992) Organometallics 11: 8
90. Guo Z, Swenson DC, Jordan RF (1994) Organometallics 13: 1424
91. Eisch JJ, Caldwell KR, Werner S, Krüger C (1991) Organometallics 10: 3417
92. Eisch JJ, Pombrik SI, Zheng G-X (1993) Organometallics 12: 3856
93. Eshuis JJW, Tan YY, Teuben JH, Renkema J (1990) J Mol Catal 62: 277
94. Hlatky GG, Turner HW, Eckman RR (1989) J Am Chem Soc 111: 2728
95. Yang X, Stern CL, Marks TJ (1994) J Am Chem Soc 116: 10015
96. Bochmann M, Lancaster SJ (1992) J Organomet Chem 434: C1
97. Bochmann M, Lancaster SJ (1993) Organometallics 12: 633
98. Taube R, Krukowa L (1988) J Organomet Chem 327: C9
99. Bochmann M, Jaggar AJ, Nicholls JC (1990) Angew Chem Int Ed Engl 29: 780
100. Chein JCW, Tsai WM, Rausch MD (1991) J Am Chem Soc 113: 8570
101. Chien JCW, Song W, Rausch M (1993) Macromolecules 26: 3239
102. Bochmann M, Wilson LM (1986) J Chem Soc, Chem Commun 1610
103. Jordan RF, Bajgur CS, Dasher WE, Rheingold AL (1987) Organometallics 6: 1041
104. Jordan RF, Dasher WE, Echols SF (1986) J Am Chem Soc 108: 1718
105. Alelyunas YW, Jordan RF, Echols SF, Borkowsky SL, Bradley PK (1991) Organometallics 10: 1406
106. Jordan RF, La Pointe RE, Bradley PK, Baenzier N (1989) Organometallics 8: 2892
107. Jordan RF, La Pointe RE, Baenziger N, Hinch GD (1990) Organometallics 9: 1539
108. Crowther DJ, Borkowsky SL, Swenson D, Meyer TY, Jordan RF (1993) Organometallics 12: 2897
109. Crasther DJ, Borkowsky SL, Swenson D, Meyer TY, Jordan RF (1993) Organometallics 12: 2897
110. Yang X, King WA, Sabat M, Marks TJ (1993) Organometallics 12: 4254
111. Thompson ME, Baxter SM, Bulls AR, Burger BJ, Nolan MC, Santarsiero BD, Schaefer WP, Bercaw JE (1987) J Am Chem Soc 109: 203
112. Burger BJ, Thompson ME, Cotter WD, Bercaw JE (1990) J Am Chem Soc 112: 1566

113. Bunel E, Burger BJ, Bercaw JE (1988) J Am Chem Soc 110: 976
114. Coughlin EB, Bercaw JE (1992) J Am Chem Soc 114: 7606
115. Shapiro PJ, Cotter WD, Schaefer WP, Labinger JA, Bercaw JE (1994) J Am Chem Soc 116: 4623
116. Hajela S, Bercaw JE (1994) Organometallics 13: 1147
117. Evans WJ, Ulibarri TA, Ziller JW (1990) J Am Chem Soc 112: 2314
118. Yang X, Stern CL, Marks TJ (1991) Organometallics 10: 840
119. Yasuda H, Ihara E (1995) Macromol Chem Phys 196: 2417
120. Schaverien CJ (1994) Organometallics 13: 69
121. Lin Z, Marechal J-FL, Sabat M, Marks TJ (1987) J Am Chem Soc 109: 4127
122. Marks TJ (1992) Acc Chem Res 25: 57
123. Huang Y, Hill YD, Sodupe M, Bauschlicher CW, Jr., Freiser BS (1992) J Am Chem Soc 114: 9106
124. Evans WJ, Meadows JH, Hunter WE, Atwood JL (1984) J Am Chem Soc 106: 1291
125. Thompson ME, Bercaw JE (1984) Pure & Appl Chem 56: 1
126. Shapiro PJ, Bunel E, Schaefer WP, Bercaw JE (1990) Organometallics 9: 867
127. Jeske G, Schock LE, Sweptson PN, Schumann H, Marks TJ (1985) J Am Chem Soc 107: 8103
128. Yasuda H, Ihara E (1994) Koubunshi 43: 534
129. Uppal JS, Janson DE, Staley RH (1981) J Am Chem Soc 103: 508
130. Christ (1988) J Am Chem Soc 110: 4038
131. Christ CS Jr., Eyler JR, Richardson DE (1990) J Am Chem Soc 112: 596
132. Jacobson DB, Freiser BS (1985) J Am Chem Soc 107: 5876
133. Huang Y, Hill YD, Freiser BS (1991) J Am Chem Soc 113: 840
134. Cossee P (1964) J Catal 3: 80
135. Arlman EJ (1964) J Catal 3: 89
136. Arlman EJ, Cossee P (1964) J Catal 3: 99
137. Doherty NM, Bercaw JE (1985) J Am Chem Soc 107: 2670
138. Tasumi M (1966) J Polym Sci, Polym Chem Ed 4: 1023
139. Ivin KH, Rooney JJ, Stewart CD, Green MLH, Mahtab R (1978) J Chem Soc, Chem Commun 604
140. Upton TH, Rappé AK (1985) J Am Chem Soc 107: 1206
141. Turner HW, Schrock RR, Fellmann JD, Holmes SJ (1983) J Am Chem Soc 105: 4942
142. Flood JC, Bitter SP (1984) J Am Chem Soc 106: 6076
143. Watson PL (1982) J Am Chem Soc 104: 337
144. Clawson L, Soto J, Buchwald SL, Steigerwald ML, Grubbs RH (1985) J Am Chem Soc 107: 3377
145. Piers WE, Bercaw JE (1990) J Am Chem Soc 112: 9406
146. Bruno JW, Smith GM, Marks TJ, Fair CK, Schultz AJ, Williams JM (1986) J Am Chem Soc 108: 40
147. Tatsumi K, Nakamura A (1987) Organometallics 6: 427
148. Tatsumi K, Nakamura A (1987) J Am Chem Soc 109: 3195
149. Kaminsky W, Ahlers A, Moller-Lindenhof N (1989) Angew Chem Int Ed Engl 28: 1216
150. Tsutsui T (1989) Polymer 30: 428
151. Guo Z, Bradley PK, Jordan RF (1992) Organometallics 11: 2690
152. Resconi L, Piemontesi F, Franciscono G, Abis L, Fiorani T (1992) J Am Chem Soc 114: 1025
153. Grassi A, Zambelli A, Resconi L, Albizzati E, Mazzocchi R (1988) Macromolecules 21: 617
154. Cheng HN, Ewen JA (1989) Macromol Chem 190: 1931
155. Chien JCW, Wang B-P (1990) J Polym Sci, Polym Chem Ed 28: 15
156. Mogstad A-L, Waymouth RM (1992) Macromolecules 25: 2282
157. Watson PL, Roe DC (1982) J Am Chem Soc 104: 6471
158. Llinas GH, Dong S-H, Mallin DT, Rausch MD, Lin Y-G, Winter HH, Chien JCW (1992) Macromolecules 25: 1242
159. Jüngling S, Mülhaupt R, Stehling U, Brintzinger H-H, Fischer D, Langhauser F (1995) J Polym Chem, Part A: Polym Chem 33: 1305
160. Spaleck W, Küber F, Winter A, Rohrmann J, Bachmann B, Antberg M, Dolle V, Paulus EF (1994) Organometallics 13: 954
161. Erker G, Nolte R, Tsay Y-H, Krüger C (1989) Angew Chem, Int Ed Engl 28: 628
162. Razavi A, Atwood JL (1993) J Am Chem Soc 115: 7529
163. Erker G, Aulbach M, Knickmeier M, Wingbermühle D, Krüger C, Nolte M, Werner S (1993) J Am Chem Soc 115: 4590

164. Erker G, Temme B (1992) J Am Chem Soc 114: 4004
165. Rieger B, Mu X, Mallin DT, Rausch MD, Chien JCW (1990) Macromolecules 23: 3559
166. Longo P, Grassi, A, Pellecchia C, Zambelli A (1987) Macromolecules 20: 1015
167. Mengele W, Diebold J, Troll C, Röll W, Brintzinger H-H (1993) Organometallics 12: 1931
168. Leclerc MK, Brintzinger HH (1995) J Am Chem Soc 117: 1651
169. Erker G (1991) Chem Ber 124: 1301
170. Erker G, Nolte R, Aul R, Wilker S, Krüger C, Noe R (1991) J Am Chem Soc 113: 7594
171. Erker G (1992) Pure and Appl Chem 64: 393
172. Erker G, Fritze C (1992) Angew Chem, Int Ed Engl 31: 199
173. Coates CW, Waymouth RM (1995) Science 267: 217
174. Cavallo L, Guerra G, Vacatello M, Corradini P (1991) Macromolecules 24: 1784
175. Corradini P, Guerra G, Vacatello M, Villani V (1988) Gazz Chim Ital 118: 173
176. Resconi L, Abis L, Franciscono G (1992) Macromolecules 25: 6814
177. Ewen JA, Elder MJ, Jones RL, Haspeslagh L, Atwood JL, Bott SG, Robinson K (1991) Makromol Chem, Macromol; Symp 48/49: 253
178. Doi Y, Ueki S, Keii T (1979) Macromolecules 12: 814
179. Doi Y, Suzuki S, Soga K (1986) Macromolecules 19: 2896
180. Ewen JA, Reddy BR, Elder MJ (1994) Eur Pat Appl 577581
181. Winter A, Rohrmann J, Dolle V, Spaleck W (1990) Ger Offen 3907965
182. Inoue N, Shiomura T, Asanuma T, Jinno M, Sonobe Y, Mizutani K (1991) Jpn Kokai Tokkyo Koho 03193793
183. Locatelli P, Sacchi MC, Rigamonti E, Zambelli A (1984) Macromolecules 17: 123
184. Ishihara N, Semiya T, Kuramoto M, Uoi M (1986) Macromolecules 19: 2464
185. Ishihara N, Kuramoto M, Uoi M (1988) Macromolecules 21: 3556
186. Kucht A, Kucht H, Barry S, Chien JCW, Rausch MD (1993) Organometallics 12: 3075
187. Pellecchia C, Proto A, Longo P, Zambelli A (1992) Makromol Chem Rapid Commun 13: 277
188. Pellecchia C, Long P, Proto A, Zambelli A (1992) Makromol Chem, Rapid Commun 13: 265
189. Quyoum R, Wang Q, Tudoret M-J, Baird MC, Gillis DJ (1994) J Am Chem Soc 116: 6435
190. Pellecchia C, Pappalardo D, Oliva L, Zambelli A (1995) J Am Chem Soc 117: 6593
191. Gillis DJ, Tudoret M, Baird MC (1993) J Am Chem Soc 115: 2543
192. Pellecchia C, Immirzi A, Grassi A, Zambelli A (1993) Organometallics 12: 4473
193. Pellecchia C, Grassi A, A Immirzi (1993) J Am Chem Soc 115: 1160
194. Pellecchia C, Grassi A, Zambelli A (1994) Organometallics 13: 298
195. Pellecchia C, Grassi A, Zambelli A (1993) J Chem Soc, Chem Commun 947
196. Pellecchia C, Longo P, Grassi A, Ammendola P, Zambelli A (1987) Macromol Chem, Rapid Commun 8: 277
197. Zambelli A, Oliva L, Pellecchia C (1989) Macromolecules 22: 2129
198. Zambelli A, Pellecchia C, Proto A (1995) Macromol Symp 89: 373
199. Grassi A, Pellecchia C, Oliva L, Laschi F (1995) Macromol Chem Phys 196: 1093
200. Flores JC, Chien JCW, Rausch MD (1995) Organometallics 14: 1827
201. Flores JC, Chien JCW, Rausch MD (1995) Organometallics 14: 2106
202. Stevens JC (1993) INSITE-CGCT, Houston Tex USA 111
203. Castonguray LA, Rappé AK (1992) J Am Chem Soc 114: 5832
204. Resconi L, Waymouth RM (1990) J Am Chem Soc 112: 4953
205. Resconi L, Coates GW, Mogstad A, Waymouth RM (1991) J Macromol Sci, Chem Ed A28 1225
206. Butler GB (1982) Acc Chem Res 15: 370
207. Young JR, Stille JR (1990) Organometallics 9: 3022
208. Cheng HN, Khaset NP (1988) J Appl Polym Sci 35: 825
209. Coates GW, Waymouth RM (1993) J Am Chem Soc 115: 91
210. Waymouth R, Pino P (1990) J Am Chem Soc 112: 4911
211. Pino P (1990) J Organomet Chem 370: 1
212. Abe H, Imai K, Matsumoto M (1965) J Polym Sci, Part B 3: 1053
213. Abe H, Imai K, Matsumoto M (1968) J Polym Sci C23: 469
214. Kesti MR, Coates GW, Waymouth RM (1992) J Am Chem Soc 114: 9679
215. Yasuda H, Yamamoto Y, Yokota K, Miyake S, Nakamura A (1992) J Am Chem Soc 114: 4908
216. Yasuda H, Yamamoto H, Yamashita M, Yokota K, Nakamura A, Miyake S, Kai Y, Kanehisa N (1993) Macromolecules 26: 7134

217. Yasuda H, Furo M, Yamamoto H, Nakamura A, Miyake S, Kibino N (1992) Macromolecules 25: 5115
218. Brokhart M (1994) J Am Chem Soc
219. Brookhart M, Sabo-Etienne S (1991) J Am Chem Soc 113: 2777
220. Hauptman E, Sabo-Etienne S, White PS, Brookhart M, Garner JM, Fagan PJ, Calabrese JC (1994) J Am Chem Soc 116: 8038
221. Boffa LS, Novak BM (1994) Macromolecules 27: 6993
222. Jiang T, Shen Q, Lin Y, Jin S (1992) J Organomet Chem 450: 121
223. Collins S, Ward DG (1992) J Am Chem Soc 114: 5460
224. Soga K, Deng H, Yano T, Shiono T (1994) Macromolecules 27: 7938
225. Deng H, Shiono T, Soga K (1995) Macromolecules 28: 3067
226. Giardello MA, Yamamoto Y, Brard L, Marks TJ (1995) J Am Chem Soc 117: 3276
227. Nomura R, Toneri T, Endo T (1994) Polymer Preprints, Japan 34: 158
228. Nakayama Y, Shibahara T, Fukumoto H, Nakamura A, Mashima K (1996) Macromolecules 29: 8014
229. Quijada R, Dupont J, Silveira DC, Miranda MSL, Scipioni RB (1995) Macromol Rapid Commun 16: 357
230. Yasuda H, Tatsumi K, Okamoto T, Mashima K, Lee K, Nakamura A, Kai Y, Kanehisa N, Kasai N (1985) J Am Chem Soc 107: 2410
231. Okamoto T, Yasuda H, Nakamura A Kai Y, Kanehisa N, Kasai N (1988) J Am Chem Soc 110: 5008
232. Okamoto T, Yasuda H, Nakamura A, Kai Y, Kanehisa N, Kasai N (1988) Organometallics 7: 2266
233. Mashima K, Yamanaka Y, Fujikawa S, Yasuda H, Nakamura A (1992) J Organomet Chem 428: C5
234. Gibson VC (1994) J Chem Soc, Dalton Trans 1607
235. Gibson VC (1994) Angew Chem, Int Ed Engl 33: 1565
236. Cockcroft JK, Gibson VC, Howard JAK, Poole AD, Siemeling U, Wilson C (1992) J Chem Soc, Chem Commun 1668
237. Buijink J-K, Teuben JH, Kooijman H, Spek AL (1994) Organometallics 13: 2922
238. Sundermeyer J, Runge D (1994) Angew Chem, Int Ed Engl 33: 1255
239. Uhrhammer R, Crowther DJ, Olson JD, Swenson DC, Jordan RF (1992) Organometallics 11: 3098
240. Williams DS, Schofield MH, Anhaus JT, Schrock RR (1990) J Am Chem Soc 112: 6728
241. Williams DS, Schofield MH, Schrock RR (1993) Organometallics 12: 4560
242. Dyer PW, Gibson VC, Howard JAK, Whittle B, Wilson C (1992) J Chem Soc, Chem Commun 1666
243. Mashima K, Tanaka Y, Nakamura A (1995) J Organomet Chem 502: 19
244. Mashima K, Tanaka Y, Nakamura A (1995) Organometallics 14: 5642
245. Mashima K, Tanaka Y, Kaidzu M, Nakamura A (1996) Organometallics 15: 2431
246. Mashima K, Fujikawa S, Nakamura A (1993) J Am Chem Soc 115: 10990
247. Mashima K, Fujikawa S, Urata H, Tanaka Y, Nakamura A (1994) J Chem Soc, Chem Commun 1623
248. Mashima K, Fujikawa S, Tanaka Y, Urata H, Oshiki T, Tanaka E, Nakamura A (1995) Organometallics 14: 2633
249. Gardiner MG, Raston CL (1993) Organometallics 12: 81
250. Mashima K, Nakamura A (unpublished results)
251. Temme B, Erker G, Karl J, Luftmann H, Fröhlich R, Kotila S (1995) Angew Chem, Int Ed Engl 34: 1755
252. Blenkers J, Hessen B, van Bolhuis F, Wagner AJ, Teuben JH (1987) Organometallics 6: 459
253. Meijer-Veldman MEE, Tan YY, de Liefde Meijer HJ (1985) Polym Commun 26: 200
254. Bazan GC, Donnely SJ, Rodriguez G (1995) J Am Chem Soc 117: 2671
255. Rodriguez G, Bazan GC (1995) J Am Chem Soc 117: 10155
256. Scheuer S, Fischer J, Kress J (1995) Organometallics 14: 2627
257. Feher FJ, Walzer JF, Blanski RL (1991) J Am Chem Soc 113: 3618
258. Feher FJ, Blanski RL (1992) J Am Chem Soc 114: 5886
259. Clark A (1969) Catal Rev 3: 145
260. Karol KFJ, Brown GL, Davision JM (1973) J Polym Sci, Polym Chem Ed 11: 413
261. Thomas BJ, H. Theopold K (1988) J Am Chem Soc 110: 5902

262. Thomas BJ, Noh SK, Sculte GK, Sendlinger SC, Theopold KH (1991) J Am Chem Soc 113: 893
263. Theopold KH (1990) Acc Chem Res 23: 263
264. Bhandari G, Kim Y, McFarland JM, Rheingold AL, Theopold KH (1995) Organometallics 14: 738
265. Coles M, Dalby CI, Gibson VC, Clegg W, Elsegood MRJ (1995) J Chem Soc, Chem Commun 1709
266. Grubbs RH, Tumas W (1989) Science 243: 907
268. Feldman J, Schrock RR (1991) Prog Inorg Chem 39: 1
269. Skell PS, Van Dam EM, Silvon MP (1974) J Am Chem Soc 96: 626
270. Skell PS, McGlinchey MJ (1975) Angew Chem, Int Ed Engl 14: 195
271. Gausing VW, Wilke G (1981) Angew Chem, Int Ed Engl 93: 201
272. Skupiñska J (1991) Chem Rev 91: 613
273. Keim W (1984) Chem Ing Tech 56: 850
274. Brookhart M, DeSimone JM, Grant BE, Tanner MJ (1995) Macromolecules 28: 5378
275. Johnson LK, Killian CM, Brookhart M (1995) J Am Chem Soc 117: 6414
276. Guillet J (1985) Polymer Photophysics and photochemistry, Cambridge University, Cambridge
277. Moore JS, Stupp SI (1992) J Am Chem Soc 114: 3429
278. van Asselt R, Gielens EECG, Rülke RE, Vrieze K, Elsevier CJ (1994) J Am Chem Soc 116: 977
279. Sen A (1993) Acc Chem Res 26: 303
280. Jiang Z, Adams SE, Sen A (1994) Macromolecules 27: 2694
281. Jiang Z, Sen A (1995) J Am Chem Soc 117: 4455
282. Brookhart M, Wagner MI, Balavoine GGA, Haddou HA (1994) J Am Chem Soc 116: 3641
283. Bronco S, Consiglio G, Hutter R, Batistini A, Suter UW (1994) Macromolecules 27: 4436
284. Nozaki K, Sato N, Takaya H (1995) J Am Chem Soc 117: 9911
285. Soga K, Kaminaka M, Kim HJ, Shiono T (1995) In Ziegler Catalysts; Fink G, Mülhaupt R and Brintzinger HH, Eds.; Springer: Berlin,: p 333
286. Soga K, Arai T, Hoang BT, Uozumi T (1995) Macromol. Rapid Commun 16: 905
287. Nishida H, Uozumi T, Arai T, Soga K (1995) Macromol Rapid Commun 16 821

Editor: T. Saegusa
Received: August 1996

Rare Earth Metal-Initiated Living Polymerizations of Polar and Nonpolar Monomers

Hajime Yasuda[1] and Eiji Ihara[2]
Department of Applied Chemistry, Faculty of Engineering,
Hiroshima University, Higashi-Hiroshima 739, Japan
[1] E-mail: yasuda@ipc.hiroshima-u.ac.jp. [2] E-mail: ihara@ipc.hiroshima-u.ac.jp.

This review article shows that by using the versatile functions of rare earth metal complexes we can polymerize both polar and nonpolar monomers in living fashion to obtain monodisperse high molecular weight polymers at high conversions. A typical example is the polymerization of methyl methacrylate with $[SmH(C_5Me_5)_2]_2$ or $LnMe(C_5Me_5)_2$ (THF) (Ln = Sm, Y, Lu), which leads quantitatively to high molecular weight syndiotactic polymers ($M_n > 500\,000$, syndiotacticity > 95%) at low temperature ($-95\,°C$). The initiation mechanism was discussed on the basis of X-ray analysis of the 1:2 adduct (molar ratio) of $[SmH(C_5Me_5)_2]_2$ with MMA. Living polymerizations of alkyl acrylates (methyl acrylate, ethyl acrylate, butyl acrylate) were also made possible by using $LnMe(C_5Me_5)_2$(THF) (Ln = Sm, Y), with the results: poly(methyl acrylate) $M_n = 48 \times 10^3$, $M_w/M_n = 1.04$, poly(ethyl acrylate) $M_n = 55 \times 10^3$, $M_w/M_n = 1.04$, and poly(butyl acrylate) $M_n = 70 \times 10^3$, $M_w/M_n = 1.05$. By taking advantage of the ABA triblock copolymerization of MMA/butyl acrylate/MMA, it was possible to obtain rubberlike elastic polymers. Lanthanum alkoxide(III) has good catalytic activity for the polymerization of alkylisocyanates ($M_n > 10^6$, $M_w/M_n = 2.08$). Monodisperse polymerization of lactones, lactide, and oxirane was also achieved by polymerization with rare earth metal complexes. C_1 symmetric bulky organolanthanide(III) complexes such as $SiMe_2[2(3),4-(SiMe_3)_2C_5H_2]_2LnCH(SiMe_3)_2$ (Ln = La, Sm, Y) show high activity for linear polymerization of ethylene. Organolanthanide(II) complexes such as racemic $SiMe_2[2-SiMe_3-4-tBu-C_5H_2]_2Sm(THF)_2$ as well as C_1 symmetric $SiMe_2[2(3),4-(SiMe_3)_2C_5H_2]_2Sm(THF)_2$ were also found to have a very high activity for polymerization of ethylene. Thus, polyethylene of $M_n > 10^6$ ($M_w/M_n = 1.60$) was obtained by using $SiMe_2[2(3),4-(SiMe_3)_2C_5H_2]_2Sm(THF)_2$. 1,4-cis Conjugated diene polymers of butadiene and isoprene became available by the efficient catalytic activity of C_5H_5NdCl/AlR_3 or $Nd(octanoate)_3/AlR_3$. The Ln(naphthenate)$_3$/AliBu$_3$ system allows selective polymerization of acetylene in cis fashion to take place at high yield. Considering the fact that rare earth metal-initiated living polymerization can be achieved for both polar and nonpolar monomers, attempts have been made to block copolymerization of ethylene with MMA or lactones yielding polyethylene derivatives having high chemical reactivity.

Advances in Polymer Science, Vol. 133
© Springer-Verlag Berlin Heidelberg 1997

List of Abbreviations

M_w	weight-average molecular weight
M_n	number-average molecular weight
MMA	methyl methacrylate
tBuMA	t-butyl methacrylate
iBuMgBr	isobutylmagnesium bromide
THF	tetrahydrofuran
MeA	methyl acrylate
EtA	ethyl acrylate
BuA	butyl acrylate
2-EP	2-ethylhexylphosphate
EO	ehtylene oxide
PO	propylene oxide
AGE	allyl glycidylether
EPH	epichlorohydrin
CL	caprolactone
VL	valerolactone
TMC	trimethylenecarbonate
P_{204}	$[CH_3(CH_2)_3(CH_2CH_3)CHCH_2O]_2P(O)OH$
P_{507}	$(iC_8H_{17}O)_2P(O)OH$
P_{215}	$[C_6H_{13}CH(CH_3)O]_2P(O)OH$
PA	polyacetylene
DSC	differential scanning calorimetry
XRD	X-ray diffraction
SEM	scanning electron micrograph
GPC	gel permeation chromatograph
v_{as}	asymmetric vibration
v_s	symmetric vibration

1 Introduction

Recently, it has been made clear that the rare earth metal-initiated living polymerizations of polar and nonpolar monomers allow the synthesis of high molecular weight polymers with a very narrow molecular weight distribution. The conditions required for ideal living polymerization may be (1) the number-average molecular weight $M_n > 100\,000$, (2) the polydispersity index $M_w/M_n < 1.05$, (3) stereoregularity $> 95\%$, and (4) high conversion in a short period of time $> 95\%$. Even the famous living polystyrene system [1] and the group transfer system [2] do not simultaneously satisfy conditions (1) and (2). Recently, after a long wait by many polymer chemists, all four conditions have been found to be met in the rare earth metal-initiated polymerization of methyl methacrylate (MMA). The success is mainly ascribed to the use of single crystalline rare earth metal complexes as initiators and that of a thoroughly dried monomer. Although single air- and moisture-sensitive complexes are generally difficult to prepare, especially for polymer chemists, it is a rather easy task for organometallic chemists who are familiar with Schlenk techniques.

The synthesis of monodisperse high molecular weight syndiotactic poly(methyl methacrylate) mentioned above was possible at a polymerization temperature as low as $-95\,^{\circ}$C [3]. However, methods to synthesize it at a relatively high temperature or to synthesize its isotactic counterpart remain to be found. When the latter is made available, the synthesis of stereo-complexes of poly(methyl methacrylate) will become the focus of considerable interest.

Alkyl acrylates were polymerized for the first time in living fashion with the aid of the unique catalytic action of rare earth metal complexes [4]. Since these monomers have an acidic α-H, termination and chain transfer reactions occur so frequently that their polymerization generally does not proceed in living manner. The lowest polydisperse index reported so far is 1.06 for poly(tBu acrylate, $M_n = 21\,000$) obtained using the RLi/LiCl system [5], 1.1 for poly(methyl acrylate, $M_n = 20\,000$) obtained by the catalysis of $Sm(C_5Me_5)_2(CH_2-CH-CH-CH_2-)_2$ [6], and 1.03 for poly(ethyl acrylate, $M_n = 3300$) obtained by the catalysis of a group transfer agent [7]. By taking advantage of the living nature of both poly(methyl methacrylate) and poly(alkyl acrylate)s, ABA-type triblock copolymerization of methyl methacrylate (MMA)/butyl acrylate(BuA)/MMA was performed to obtain thermoplastic elasticity. Since MMA acts as a hard segment and BuA as a soft segment, the former will work as a node and the latter as an elastomer in the resulting network system. In fact, the poly(MMA/BuA/MMA)(8:72:20) system obtained exhibited good elastic properties [4], while the (20:47:33, 6:91:3) system did not. How syndiotactic or isotactic poly(alkyl acrylate) with relatively large M_n and relatively small M_w/M_n can be produced remains to be elucidated.

Living polymerization of lactones has been successfully performed by catalysis of rare earth metal complexes to obtain M_w/M_n of 1.07–1.08 [8]. The lowest polydispersity index attained so far with the $AlEt_3/H_2O$ system is 1.13.

Dielectric normal mode relaxation has been examined using the resulting pure poly(lactone) [9]. Cyclic carbonates such as trimethylenecarbonate also undergo living polymerization [10]. However, it should be noted that their complete drying was mandatory for the synthesis of high molecular weight polymers. More recently, acrylonitrile and alkyl isocyanates have been polymerized using $(C_5Me_5) LaCH(SiMe_3)_2$. The polymerization of various oxiranes has also been attempted using the $Ln(acac)_3/AlR_3/H_2O$ system [11]. All these recent studies have made it clear that most polar monomers can be polymerized by using various organolanthanide catalysts. There are several polymerization systems which do not satisfy the four conditions mentioned earlier. The use of a single-component catalyst of suitable ligand bulkiness is necessary to achieve the ideal living polymerization of propylene oxide, propylene sulfide, and ethylene oxide.

Rare earth metal initiators also show good activity towards nonpolar monomers such as ethylene, 1-olefins, styrene, conjugated diene, and acetylene. In fact, the polymerization of poly(ethylene) of $M_n > 1\,000\,000$ with $M_w/M_n = 1.6$ was made possible by the use of a C_1 symmetric complex, $Me_2Si[3,4-(SiMe_3)_2C_5H_2]$ $[2,4-(SiMe_3)_2C_5H_2]Sm(THF)$. This M_w/M_n value is the smallest ever obtained for poly(ethylene). The conventional $MgCl_2$-assisted Ziegler-Natta catalyst produced a polymer of $M_w/M_n = 4.0$, and the Kaminsky-type catalyst a polymer of $M_w/M_n = 2.0$. However, the magnitudes of M_n for poly(olefin) s such as poly(1-pentene), polystyrene, and poly(1,5-hexadiene) did not reach high enough levels even with the help of the characteristic catalytic action of rare earth metal complexes. The maximum $M_n = 60\,000$ obtained is far below the desired minimum of 400 000. Furthermore, the synthesis of syndiotactic poly(1-olefin) is desirable because excellent physical properties are expected. Conjugate diene and acetylene derivatives were polymerized by using rather complex rare earth metal catalysts, but the M_w/M_n obtained remained fairly large. This indicates that the polymerization should be performed using multisite catalytic species. The use of a mononuclear single-site catalyst is required for the polymerization of these derivatives.

2 Rare Earth Metal-Initiated Polymerization of Polar Vinyl Monomers

2.1 General Remarks

Rare earth elements are the general term for 15 kinds of lanthanide elements (La, Ce, Pr, Nd, Pm, Sm, Eu, Gd, Tb, Py, Ho, Er, Tm, Yb, Lu) together with Sc and Y elements. They prefer trivalent states in the complex formation, though three elements (Eu, Sm, Yb) can assume tri- and divalent states and Ce a tri- or tetravalent state. Their ionic radii are fairly large (1.0–1.17 Å) and their electronegativities are low (1.1–1.2). In fact, the former are much larger than those of the main group metals (Li 0.73; Mg 0.71; Al 0.68 Å) and transition metals (Ti 0.75; Mo 0.75; Fe 0.75 Å), while the latter are similar to those of Li (1.0) and Mg (1.2) (Table 1). With regard to these properties, the organometallic complexes of rare earth metals have strong basicity and therefore allow anionic polymerization to take place. Rare earth metal complexes often form -ate complexes such as R_4LnLi and $R_2LnClNaCl(THF)_2$, but it was possible to obtain neutral complexes by using a bulky C_5Me_5 or a C_8H_8 ligand. Characteristic reactions were observed in their reaction with CO. For example when it reacts with CO, $(C_5H_5) Lu[C(CH_3)_3](THF)$ produces acyl complexes, leading to a double insertion complex of CO by reaction with two equivalent CO [12] Scheme 1. The double insertion of the CO molecule into the C=C [13] or N=N double bond [14] was also observed for a rare earth metal.

Characteristic initiation behavior of rare earth metals was also found in the polymerization of polar and nonpolar monomers. In spite of the accelarated development of living isotactic [15] and syndiotactic [16] polymerizations of methyl methacrylate (MMA), the lowest polydispersity indices obtained remain in the region of $M_w/M_n = 1.08$ for an M_n of only 21 200. Thus, the synthesis of high molecular weight polymers $(M_n > 100 \times 10^3)$ with $M_w/M_n < 1.05$ is still an important target in both polar and nonpolar polymer chemistry. Undoubtedly, the availability of compositionally pure materials is a must for the accurate physical and chemical characterization of polymeric materials.

Table 1. Ionic radii and electronegativities of rare earth elements

Elements	Ionic radius (Å)	Electro-negativities	Elements	Ionic radius (Å)	Electro-negativities
Sc	0.89		Li	0.73	1.0
Y	1.04		Mg	0.71	1.2
La	1.17		Al	0.68	1.5
Sm	1.11	1.1–1.2	Ti	0.75	1.5
Eu	1.09		Mo	0.75	1.8
Yb	1.01		Fe	0.75	1.8
Lu	1.00		Pt	0.94	2.2

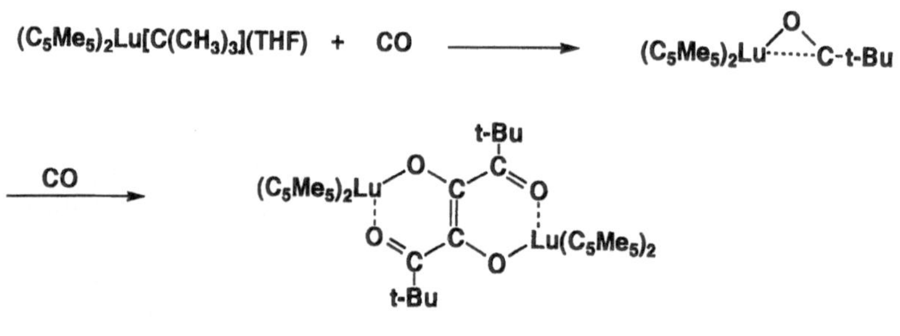

Scheme 1. Typical reactions of alkyllanthanides with CO

2.2 Highly Stereospecific Living Polymerization of Alkyl Methacrylates

As mentioned above, it is very important to find ways in which to synthesize highly syndiotactic or isotactic polymers with $M_n > 500\,000$ and $M_w/M_n < 1.05$. Although various living polymerization systems have been proposed for this purpose, anionic [17], cationic [18], group transfer [2], and metal carbene-initiated polymerizations [19] have not been successful. Therefore, it is remarkable that in recent years, high molecular weight poly(MMA) having an unusually low polydispersity has been synthesized by using the unique initiating function of organolanthanide (III) complexes (Scheme 2, Fig. 1) [3]. The relevant complexes include lanthanide hydrides, trialkylaluminum

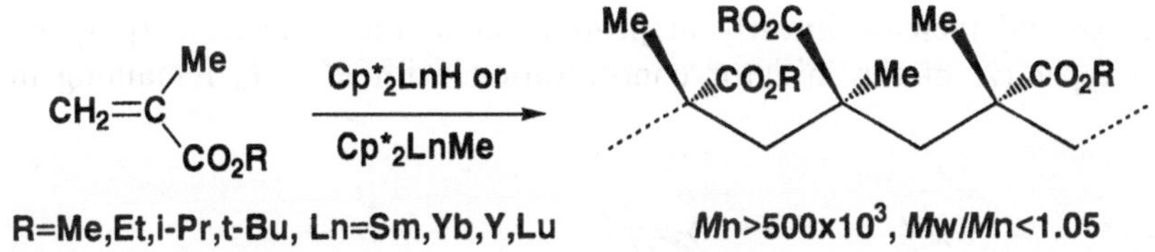

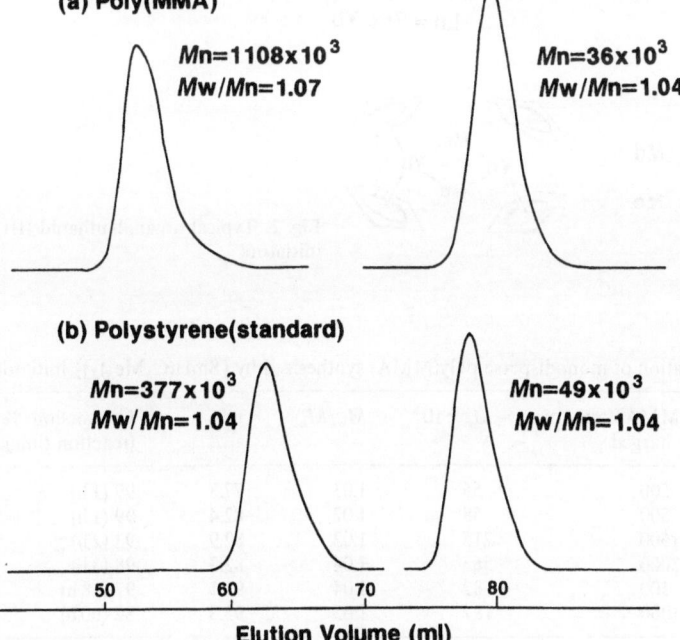

Scheme 2. Living polymerization of MMA

Fig. 1. GPC profiles of poly(MMA) and standard poly(styrene)

complexes of alkyllanthanides, balky alkyllanthamide and simple alkyl complexes synthesized from $Ln(C_5Me_5)_2Cl$ (Fig. 2). Most of them have been characterized by single X-ray analysis [20]. The results from the polymerization of MMA with the $[SmH(C_5Me_5)_2]_2$ initiator at different temperatures are summarized in Table 2. The most striking is $M_w/M_n = 1.02-1.04$, for $M_n > 60 \times 10^3$. Remarkably, the organolanthanide complexes give high conversions (polymer yields) in a relatively short period and allow the polymerization to proceed over a wide range of reaction temperatures (from -78 to $60\,°C$). Furthermore, syndiotacticity exceeding 95% is obtained when the polymerization temperature is lowered to $-95\,°C$.

From these findings we see that the ideal living polymerization is made possible when $[SmH(C_5Me_5)_2]_2$ or $LnR(C_5Me_5)_2$ (Ln = Y, Sm, Yb, Lu) is used as the initiator. A polymerization process can be judged as living if the following four criteria are met. First, the number-average molecular weight of the resulting polymer should increase linearly in proportion to the conversion (polymer yield) irrespective of the initiator concentration, with M_w/M_n remaining in

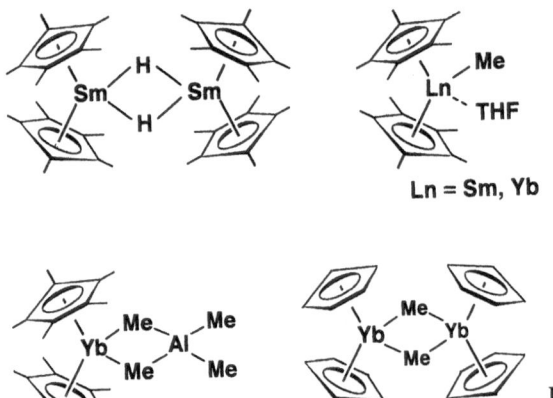

Ln = Sm, Yb

Fig. 2. Typical organolanthanide(III) initiators

Table 2. Characterization of monodisperse poly(MMA) synthesized by $[SmH(C_5Me_5)_2]_2$ initiator

Polymerization temperature (°C)	MMA/initiator charged	$M_n \times 10^3$	M_w/M_n	rr%	Conversion/% (reaction time)
40	500	55	1.03	77.3	99 (1 h)
0	500	58	1.02	82.4	99 (1 h)
0	1500	215	1.03	82.9	93 (2 h)
0	3000	563	1.04	82.3	98 (3 h)
−78	500	82	1.04	93.1	97 (18 h)
−95	1000	187	1.05	95.3	82 (60 h)

M_n and M_w/M_n were determined by GPC using standard poly(MMA) whose M_w was measured by light scattering method. Solvent, toluene; $[M]_0$/solvent = 5 (vol/vol)

the 1.03–1.04 range during the reaction (Fig. 3). Second, log $[M]_0/[M]_t$ varies linearly with time, as the result of constancy of the number of growing polymer ends. Third, the maximum efficiency of the initiator should be greater than 95%; actually, neither termination nor chain transfer should take place. Fourth, homogeneous polymerization proceeds very rapidly and is completed within 30 min when 200 equivalents of monomer are fed to the initiation system.

In general, syndiotacticity (rr%) increases with a reduction of the polymerization temperature. In the case of $SmH(C_5Me_5)_2$, it increased from 78 to 95.2% as the polymerization temperature was reduced from 25 to $-95\,°C$, but the polydispersity index remained low [3]. Extrapolating the data suggests that syndiotacticity over 97% may be obtained at $-115°C$. Polymerization of MMA in both THF and toluene using the organolanthanide initiators produced syndiotactic polymers, despite the fact that the RMgX initiator in toluene led to isotactic polymers [15].

The typical initiator systems reported so far for the synthesis of highly syndiotactic poly(MMA) are bulky alkyllithium $CH_3(CH_2)_4CPh_2Li$ [21], Grignard reagents in THF [22], and some AlR_3 complexes [23]. Although the first initiator in THF reacted rapidly with MMA at $-78\,°C$, the M_n reached only 10 000 with $M_w/M_n = 1.18$, while it produced isotactic polymers in toluene. Isobutylmagnesium bromide and vinylbenzylmagnesium bromide in THF at lower temperatures also produced high syndiotacticity, but M_n remained as low as 14 000–18 000 and the yields were quite low. When iBuMgBr or tBuMgBr

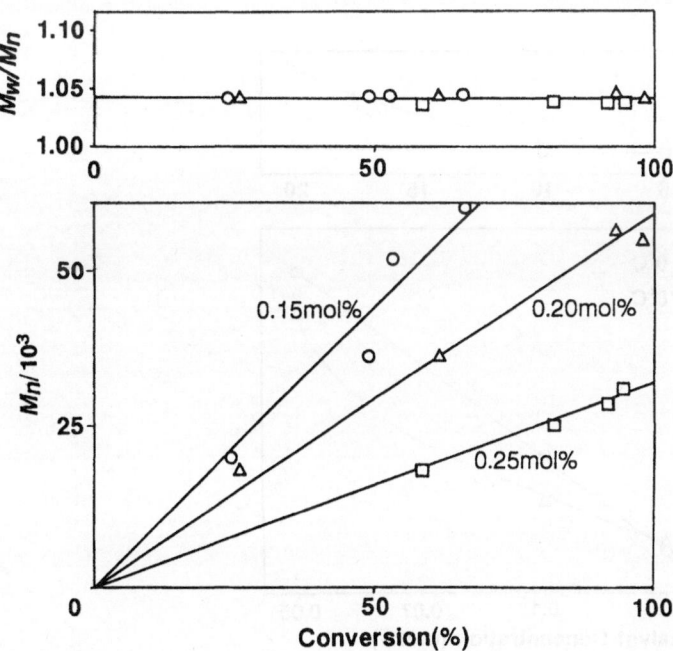

Fig. 3. M_w/M_n and M_n vs. conversion plots for poly(MMA) obtained using $[SmH(C_5Me_5)_2]_2$

was used in toluene instead of THF, the resulting poly(MMA) had a high isotacticity of 96.7% with $M_n = 19\,900$ and $M_w/M_n = 1.08$ [24]. $AlEt_3 \cdot PR_3$ complexes produced a high syndiotacticity, but not a high molecular weight [23].

Ketene silyl acetal/nucleophilic agent systems initiate the polymerization of alkyl methacrylates. These well-known group transfer systems yielded living polymers with atactic sequences at relatively high temperatures [2]. $Me_2C=C(OMe)OSiR_3$ and $R_2POSiMe_3$ can be used as initiators, and tris(dimethylamino) sulfonium bifluoride and Et_4NCN are frequently used as catalysts. For example, the M_w/M_n of the resulting poly(MMA) was 1.06 for $M_n = 3800$, and 1.15 for $M_n = 6300$. In the polymerization of butyl acrylate, the molecular weight obtained was only 4800 with an M_w/M_n of 2.14. Thus, we conclude that organolanthanide-initiated polymerization is superior in obtaining monodisperse high molecular weight poly(MMA) and poly(alkyl acrylate). However, the group transfer systems have some advantages in synthesizing poly(acrylonitrile) and poly(N,N-dimethylacrylamide), which are hardly produced with conventional organolanthanide initiators.

For the organolanthanide system, which allows the living polymerization of MMA, Yasuda et al. [3] studied the dependence of M_n on the initiator concentration (Fig. 4). When the reaction was conducted at $0\,°C$, an M_n of 530×10^3 with $M_w/M_n = 1.04$ was obtained quantitatively at an initiator concentration of $0.05\,mol\%$. This M_n value is the largest ever reported for living polymers. However, when the reaction temperature was lowered to $-78\,°C$, M_w/M_n

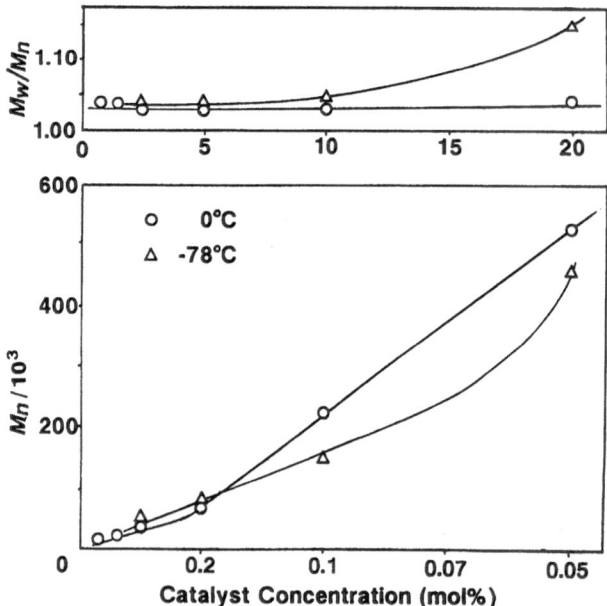

Fig. 4. M_n and M_w/M_n vs. catalyst concentration plots, initiator $[SmH(C_5Me_5)_2]_2$

increased to 1.15, due presumably to the occurrence of undesirable side reactions with water or oxygen contaminants in the monomer or solvent. In order to evaluate the living nature of the polymerization, 160 equivalents of MMA were first polymerized and a polymer of $M_n = 16 \times 10^3$ was obtained at a quantitative yield. After 30–40 min, 160 equivalents of MMA were added further to the system, which was then maintained at 0 °C. Living polymerization took place after 1 h and produced poly(MMA) of $M_n = 32 \times 10^3$ with $M_w/M_n = 1.03$ at 99% yield. The polymer produced after a 10-h reaction had an $M_n = 53 \times 10^3$ with $M_w/M_n = 1.03$, and the yield was still high (78%), indicating that only 22% of the growing ends were "dead" or dormant. Interestingly, such a long life of growing polymer ends has never been observed for other polymeric systems. When tBuMgBr- or iBuMgBr-initiated polymerization was carried out at 0 °C, the resulting poly(MMA) showed broad polydispersity ($M_w/M_n > 2.5$). This implies the need of a lower temperature for the living polymerization using the RMgX system.

These findings motivated Yasuda and coworkers [3] to isolate the 1:1 or 1:2 adduct of $[SmH(C_5Me_5)_2]_2$ with MMA in order to elucidate the initiation mechanism. They obtained the desired 1:2 adduct as an air-sensitive orange crystal (mp 132 °C). Upon deuterolysis it produced $DCMe(CO_2Me)$ $CH_2CMe(CO_2Me)CH_3$ at an excellent yield (90%), suggesting that it contains either a Sm-enolate or a Sm α-C bond (Scheme 3). The addition of 100 equivalents of MMA yielded poly(MMA) with $M_n = $ ca. 11×10^3 and $M_w/M_n = 1.03$. This result provides evidence that the 1:2 adduct is a real active species. In this way, Yasuda et al. [3] succeeded for the first time in isolating the intermediate and were able to prove that the initiation process occurs much faster than the propagation process.

Yasuda et al. [3c] also examined the eight-membered structure of $Sm(C_5Me_5)_2(MMA)_2H$ by single X-ray analysis (Rw = 0.082) (Fig. 5), finding that one of the MMA units is linked to the metal in an enolate form and at the other end the penultimate MMA unit is attached to the metal through its C=O group. Although similar cyclic intermediates have been proposed by Bawn et al. [25] and Cram and Kopecky [26] for the isotactic polymerization of MMA, the isolation of such active species has not been successful. It is expected that, in the initiation step, the hydride attacks the CH_2 group of MMA, and a transient $SmOC(OCH_3)=C(CH_3)_2$ species is formed. Then, the incoming MMA molecule is assumed to participate in the 1,4-addition to produce an eight membered cyclic intermediate (Fig. 6). Further addition of MMA to

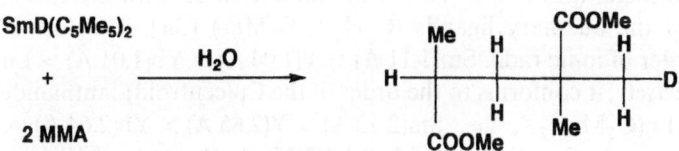

Scheme 3. Hydrolysis product of the 1:2 adduct

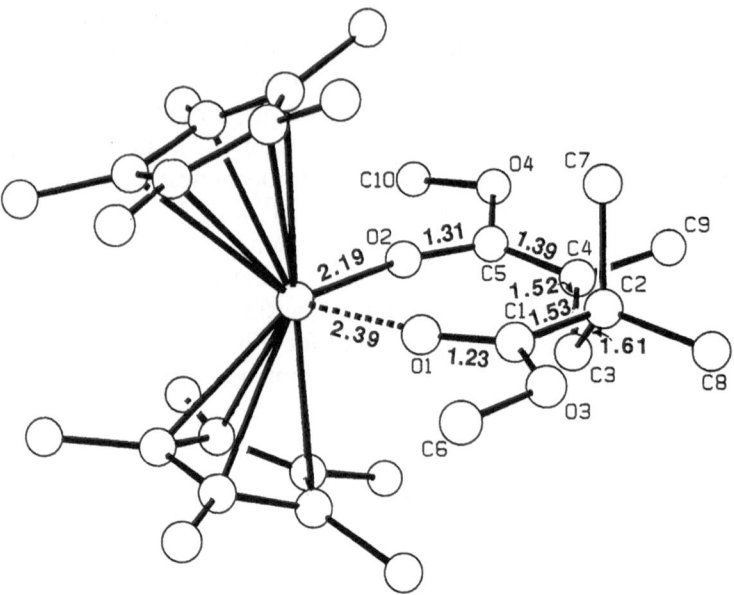

Fig. 5. X-ray structure of $Sm(C_5Me_5)_2(MMA)_2H$

the 1:2 addition compound liberates the coordinated ester group and the eight membered cyclic intermediate is again generated. In order to ascertain the above-mentioned initiation mechanism, Yasuda et al. [3c] allowed $[Y(C_5Me_5)_2OCH=CH_2]_2$ to react with excess MMA (100 equimolars: monomer/initiator = 100 mol/mol) and obtained poly(MMA) with $M_n = 10\,200$ and $M_w/M_n = 1.03$ at high yield. However, $[Y(C_5Me_5)_2OMe]_2$ and $Sm(C_5Me_5)_2OEt(Et_2O)$ were completely inert in the polymerization of MMA. Other related rare earth metal complexes such as $YMe(C_5Me_5)_2(OEt_2)$, $YbMe(C_5Me_5)_2(THF)$, $[YMe(C_5H_5)_2]_2$, $SmMe_2AlMe_2(C_5Me_5)_2$, SmMe $(C_5Me_5)_2(THF)$, and $LuMe(C_5Me_5)_2(THF)$ also showed comparably high initiating activity, and they all allowed living syndiotactic polymerization of MMA, producing polymers with $M_n = 65-120 \times 10^3$ and $M_w/M_n = 1.03-1.05$. he efficiency of these initiators was higher than 90%. Lowering the reaction temperature to $-78\,°C$ led to the production of syndiotactic polymers with rr > 90% in all cases.

The apparent rate of polymerization (the activity) increased with increasing ionic radius of the metal (Sm > Y > Yb > Lu) and decreased with increasing steric bulkiness of the auxiliary ligands ($C_5H_5 > C_5Me_5$) [3a]. The former agrees with the order of ionic radii, Sm(1.11 Å) > Y(1.04 Å) > Yb(1.01 Å) > Lu (1.00 Å). More precisely, it conforms to the order of the Cp(centroid)-lanthanide bond length of $Ln(C_5Me_5)_2X$, i.e., Sm(2.72 Å) > Y(2.65 Å) > Yb(2.64 Å) > Lu(2.63 Å). This suggests that the use of $LnMe(C_5Me_5)_2$ (Ln = La, 1.17 Å) is preferable to obtain high molecular weight polymers in a short time.

On the basis of the X-ray structural data as well as the mode of polymerization, Yasuda et al. [3a] proposed a coordination anionic mechanism involving an eight membered transition state for the organolanthanide-initiated polymerization of MMA (Fig. 6). The steric control of the polymerization reaction may be ascribed to the intermolecular repulsion between C(7) and C(9) (or the polymer chain), since completely atactic polymerization occurred when the monomer was methyl or ethyl acrylate.

Recently, isotactic polymerization of MMA has also been achieved (mm = 94%, $M_n = 134 \times 10^3$, $M_w/M_n = 6.7$) by using $Me_2Si(C_5Me_4)(C_5H_3$-1S,2S,5R-neomenthyl)LaR [R = $CH(Me_3Si)_2$ or $N(Me_3Si)_2$] [27], but $Me_2Si(C_5Me_4)(C_5H_3$-1S,2S,5R-menthyl) LnR [Ln = Lu, Sm; R = $CH(Me_3Si)_2$ or $N(Me_3Si)_2$] has been effective in producing a syndiotactic poly(MMA) (rr = 69%, $M_n = 177 \times 10^3$, $M_w/M_n = 15.7$) (Fig. 7). A possible explanation for this difference is that the menthyl complex produces a syndiotactic polymer via a cycle eight membered intermediate, while the neomenthyl complex produces an isotactic polymer via a noncyclic intermediate. To confirm this concept, iPr(Cp) (2,7-tBu-fluorenyl)YCH(Me_3Si)_2 was synthesized and used to isotactically polymerize MMA [28]. The resulting polymer had 78–79% syndiotacticity, contrary to expectation. Therefore, it was concluded that stereoregularity varies with a subtle difference in steric bulkiness between the complexes. Actually, isotactic poly(MMA) (mm = 97%) was obtained ($M_n = 200,000$, $M_w/M_n = 1.11$) when the nonmetallocene system, $[(Me_3Si)_3C]_2Yb$, was used [29].

Boff and Novac [6] found a divalent rare earth metal complex, $(C_5Me_5)_2Sm$, to be a good catalyst for the polymerization of MMA. The initiation started with

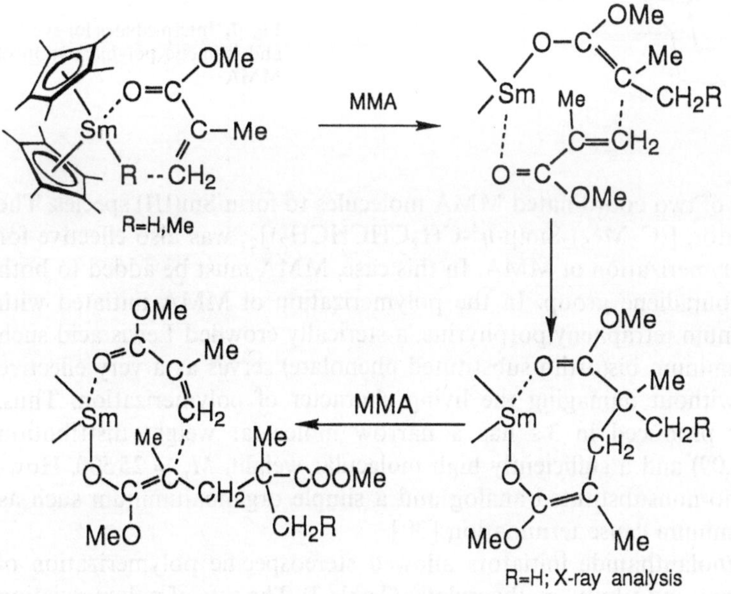

Fig. 6. Initiation mechanism for polymerization of MMA

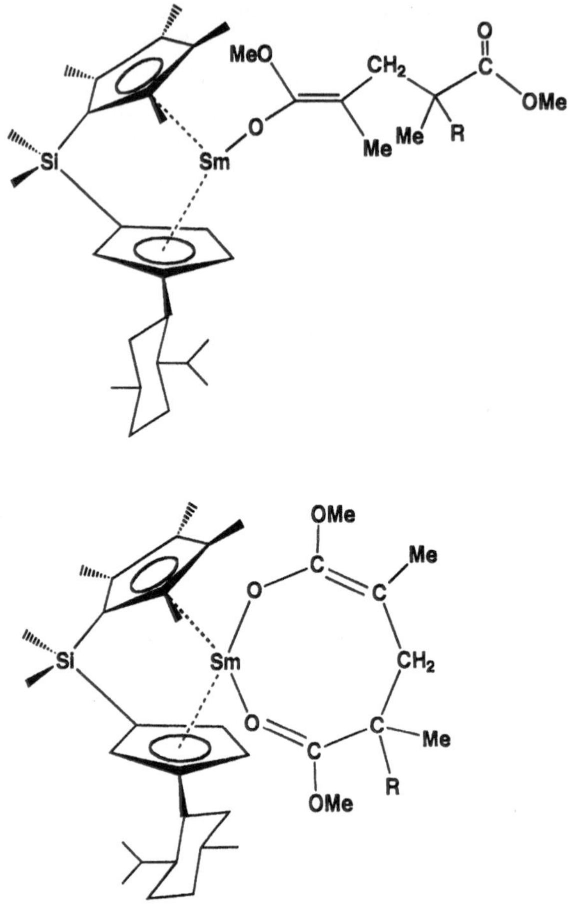

Fig. 7. Intermediates for syndio- and isotactic polymerization of MMA

the coupling of two coordinated MMA molecules to form Sm(III) species. The bisallyl initiator, $[(C_5Me_5)_2Sm(\mu-\eta^3-CH_2CHCHCH_2)]_2$, was also effective for the living polymerization of MMA. In this case, MMA must be added to both ends of the butadiene group. In the polymerization of MMA initiated with methylaluminum tetraphenylporphyrine, a sterically crowded Lewis acid such as methylaluminum bis(ortho-substituted phenolate) serves as a very effective accelerator without damaging the living character of polymerization. Thus, the polymer produced in 3 s has a narrow molecular weight distribution ($M_w/M_n = 1.09$) and a sufficiently high molecular weight, $M_n = 25\,500$. However, an ortho-nonsubstituted analog and a simple organoaluminum such as trimethylaluminum cause termination [30].

The organolanthanide initiators allowed stereospecific polymerization of ethyl, isopropyl, and t-butyl methacrylates (Table 3). The rate of polymerization and the syndiotacticity decreased with increasing bulkiness of the alkyl group in the order Me > Et > iPr = tBu. Butyl methacrylate was also polymerized using Nd(octanoate)$_3$/AlisoBu$_3$ (Al/Nd = 7–10), but the molecular weight distribution and stereoregularity were not reported [31].

In general, Ziegler-Natta catalysts such as TiCl$_4$/MgCl$_2$/AlR$_3$ and Kaminsky catalysts such as Cp$_2$ZrCl$_2$/(AlMe$_2$-O-)$_n$ do not catalyze the polymerization of polar monomers. However, a mixture of cationic species Cp$_2$ZrMe(THF)$^+$ and Cp$_2$ZrMe$_2$ has been found to do so for MMA [32], allowing syndiotactic poly(MMA) (rr = 80%, M_n = 120 000, M_w/M_n = 1.2–1.3) to be obtained. Recently, Soga et al. [33] reported the syndio-rich polymerization of MMA catalyzed by Cp$_2$ZrMe$_2$/B(C$_6$F$_5$)$_3$ or Ph$_3$CB(C$_6$F$_5$)$_4$/ZnEt$_2$ and also the isotactic polymerization of MMA catalyzed by rac-Et(ind)$_2$ZrMe$_2$/B(C$_6$F$_5$)$_3$/ZnEt$_2$ (Scheme 4).

Table 3. Organolanthanide-initiated polymerization of alkyl methacrylates

Initiator	Monomer	$M_n/10^3$	M_w/M_n	rr %	Conversion/%
[SmH(C$_5$Me$_5$)$_2$]$_2$	MMA	57	1.03	82.4	98
	EtMA	80	1.03	80.9	98
	iPrMA	70	1.03	77.3	90
	tBuMA	63	1.42	77.5	30
LuMe(C$_5$Me$_5$)$_2$(THF)	MMA	61	1.03	83.7	98
	EtMA	55	1.03	81.0	64
	iPrMA	42	1.05	80.0	63
	tBuMA	52	1.53	79.5	20

Polymerization conditions: 0°C in toluene, initiator concentration 0.2 mol%

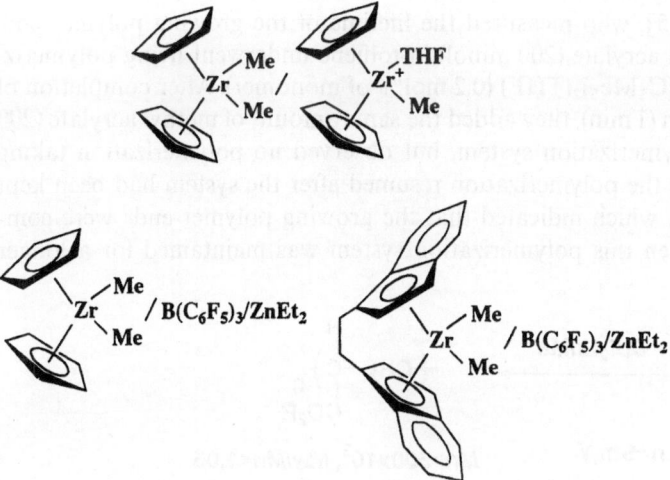

Scheme 4. Zirconium catalytic systems useful for polymerization of MMA

2.3 Living Polymerization of Alkyl Acrylates

In general, living polymerization of alkyl acrylates is difficult because chain transfer or termination occurs owing to a high sensitivity of the acidic α-proton to nucleophilic attack. Exceptions are the living polymerization of a bulky acrylic ester catalyzed by the bulky alkyllithium/inorganic salt (LiCl) system [5] as well as the group transfer polymerization of ethyl acrylate with ZnI_2 as the catalyst [2]. Aluminum-porphyrin initiator systems also induced the living polymerization of t-butyl acrylate [34], but the upper limit of molecular weight attained was ca. 20 000.

Ihara et al. [35] found the efficient initiating properties of $SmMe(C_5Me_5)_2(THF)$ and $YMe(C_5Me_5)_2(THF)$ for living polymerization of acrylic esters, i.e., methyl acrylate (MeA), ethyl acrylate (EtA), and butyl acrylate (BuA) (Scheme 5), although the reactions were nonstereospecific (Table 4). The initiator efficiency exceeded 90% except for tBuA. Ihara et al. [35] therefore concluded that the reactions occur in living fashion. In fact, the M_n of poly(BuA) initiated by $SmMe(C_5Me_5)_2(THF)$ increased linearly in proportion to the conversion, while M_w/M_n remained unchanged, irrespective of the initiator concentration. The rate of polymerization increased with increasing bulkiness of alkyl acrylates in the order nBuA > EtA > MeA, but the order was reversed in the case of methacrylic ester systems, MMA > EtMA (ethyl methacrylate) > BuMA(butyl methacrylate). For MeA, the polymerization was completed in 300 s, and both M_n and conversion increased with the polymerization time. On the other hand, the polymerization was completed in 5 s for EtA and BuA. When the initiator concentration was decreased from 0.1 to 0.02 mol%, high molecular weight poly(EtA) of $M_n = 400 000$ with a narrow molecular weight distribution ($M_w/M_n = 1.05$) was obtained, showing the characteristic nature of these initiation systems. In a similar manner, high molecular weight poly(BuA) and poly(MeA) were synthesized successfully.

Ihara et al. [35], who measured the lifetime of the growing polymer end, found that methyl acrylate (200 mmol) in toluene underwent living polymerization with $SmMe(C_5Me_5)_2(THF)$ (0.2 mol% of monomer). After completion of the polymerization (1 min), they added the same amount of methyl acrylate (200 mmol) to the polymerization system, but observed no polymerization taking place. In contrast, the polymerization resumed after the system had been kept for 1 min in THF, which indicated that the growing polymer ends were completely living. When this polymerization system was maintained for a longer

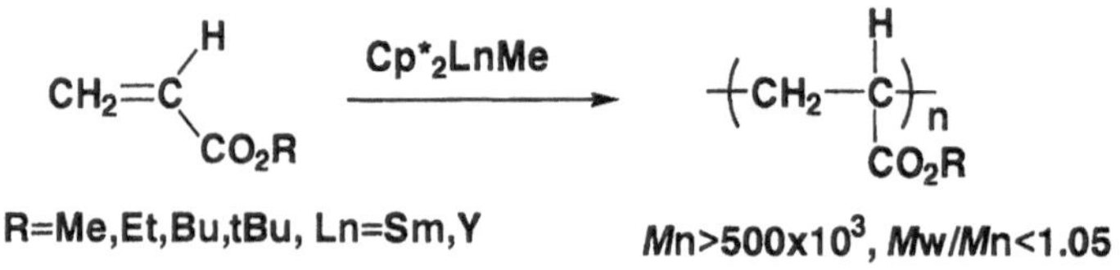

$$CH_2=C\overset{H}{\underset{CO_2R}{}} \xrightarrow{Cp^*_2LnMe} \left(CH_2-\overset{H}{\underset{CO_2R}{C}}\right)_n$$

R=Me,Et,Bu,tBu, Ln=Sm,Y $Mn>500\times10^3$, $Mw/Mn<1.05$

Scheme 5. Living polymerization of alkyl acrylates

Table 4. Polymerization of alkyl acrylates initiated by organolanthanide complexes

Initiator	Monomer	$M_n/10^3$	M_w/M_n	Tacticity/%				Conversion %	Initiator efficiency/%
				rr	mr		mm		
SmMe(C$_5$Me$_5$)$_2$(THF)	MeA	48	1.04	30	50		20	99	89
	EtA	55	1.04			51		94	86
	BuA	70	1.05	28	53		19	99	91
	tBuA	15	1.03	27	47	49	26	99	79
YMe(C$_5$Me$_5$)$_2$(THF)	MeA	50	1.07	33	51		16	99	86
	EtA	53	1.05			47		96	91
	BuA	72	1.04	22	51		27	98	88
	tBuA	17	1.03	25	45	53	30	99	75

Polymerization conditions: 0°C in toluene, initiator concentration 0.2 mol%

time (10 min at 0 °C) all the growing polymer ends were dead. Thus, the final polymer formed had a rather broad molecular weight distribution.

To gain insight into the initiation mechanism, Ihara et al. [35] examined the 1:1 reactions of $SmMe(C_5Me_5)_2$ with methyl acrylate and t butyl acrylate. Quenching of the reaction mixture of methyl acrylate and organolanthanide yielded $CH_3CH_2CH(CO_2Me)CH_2CH(CO_2Me)$ H at 65% yield. More efficient initiating properties were found when t butyl acrylate was allowed to react with $SmMe(C_5Me_5)_2(THF)$, yielding $CH_3CH_2CH(CO_2tBu)CH_2CH(CO_2tBu)$ H to 96% at 0 °C upon hydrolysis. Thus, the constitution of the acrylic ester dimer is essentially the same as that observed with the addition of MMA to $SmH(C_5Me_5)_2$. Therefore, the formation of an eight-membered intermediate is highly probable in this case too.

Ihara et al. [35] measured the tacticity at temperatures from -78 to $60\,°C$, but observed little change. This may be ascribed to the absence of steric repulsion between H(7) and H(9) atoms. The organolanthanide system was found to initiate the living copolymerization of methyl acrylate with ethyl or butyl acrylate and also that of ethyl acrylate with butyl acrylate. Table 5 shows the monomer reactivity ratio determined by using the Fineman-Ross equation. It is seen that the monomer reactivity ratio of methyl acrylate is higher than that of ethyl acrylate, but similar to that of n butyl acrylate. By contrast, the monomer reactivity ratios of alkyl acrylates are higher than that of MMA. Although in free radical initiating systems using AIBN, MMA exhibits a higher reactivity than methyl acrylate, i.e., $r_1(MMA) = 1.0$, $r_2(MeA) = 0.5$, the reverse is the case in the $NaNH_2$ initiating system, $r_1(MMA) = 0.1$ and $r_2(MeA) = 4.5$ [35c, d]. This implies that block copolymerization occurs preferentially at high conversion in the case of lanthamide initiator even when a mixture of alkyl acrylate and methyl methacrylate is used. In fact, in the random copolymerization of MMA with BuA (charged ration 1:1), the BuA component in the resulting copolymer is higher than 88% when the conversion reaches 50%.

ABA-type triblock copolymerization of MMA/BuA/MMA should give rubberlike elastic polymers. The resulting copolymers should have two vitreous outer blocks, where the poly(MMA) moiety (hard segment) associates with the nodules, and the central soft poly(BuA) elastomeric block provides rubber elasticity. Ihara et al. [35] were the first to synthesize an AB-type block copolymer, with MMA (190 equivalents of initiator) first polymerized by

Table 5. Monomer reactivity ratios of alkyl acrylates and MMA

	r_1	r_2	$r_1 \cdot r_2$
MeA/EtA	0.959	0.597	0.573
MeA/nBuA	0.426	0.578	0.246
EtA/nBuA	0.535	0.834	0.466
MMA/MeA	0.0145	19.9	0.289
MMA/EtA	0.0084	15.9	0.133
MMA/nBuA	0.0048	42.6	0.205

Initiator $SmMe(C_5Me_5)_2(THF)$ in toluene at 0 °C

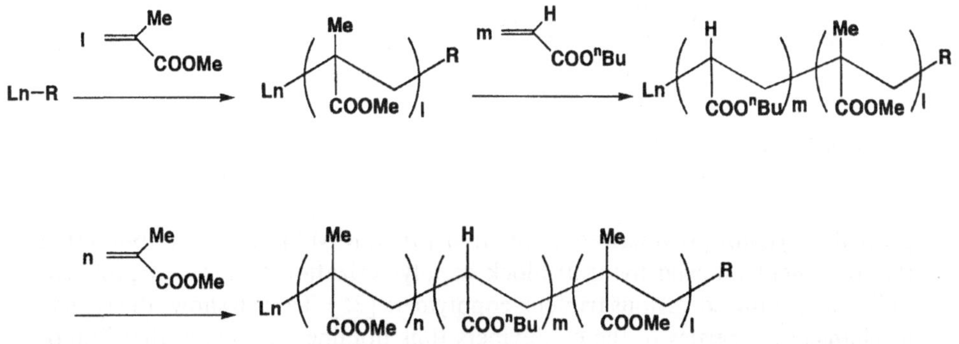

Scheme 6. ABA triblock copolymerization of MMA/butyl acrylate/MMA

Table 6. Mechanical properties of triblock copolymers

Copolymer	Tensile modulus (MPa)	Tensile strength (MPa)	Elongation (%)	Izod impact strength (J/m)	Compression set/% (70 °C, 22 h)
Poly(MMA)	610	80	21	18	100
poly(MMA/BuA/MMA) (20:47:33)	75	27	83	383 (N.B.)	101
poly(MMA/BuA/MMA) (25:51:24)	46	22	81	390 (N.B.)	103
poly(MMA/BuA/MMA) (8:72)	0.8	0.7	163	400 (N.B.)	58
poly(MMA/BuA/MMA) (6:91:3)	0.2	0.1	246	410 (N.B.)	97
poly(MMA/EtA/MMA) (26:48:26)	119	22	276	34	62

N.B., not break

SmMe$(C_5Me_5)_2$(THF) and then BuA (290 equivalents of initiator) polymerized to the poly(MMA). It consisted of 52 MMA and 48 BuA with $M_n = 47\,000$ and $M_w/M_n = 1.04$. When the BuA was first polymerized and MMA was polymerized after the system had been maintained for 5 min, the resulting polymer had a lower molecular weight ($M_n = 21\,700$) and a higher polydispersity ($M_w/M_n = 1.19$). This result may be due to partial deactivation of the living polymer chain end during the maintaining the system at ambient temperature. Thus, BuA was added immediately after MMA was polymerized for 30 min; after BuA was polymerized for 2 min, MMA was again added (Scheme 6). The first step produced poly(MMA) with $M_n = 15\,000$ and $M_w/M_n = 1.04$, the second step produced poly(MMA-blo-BuA) with $M_n = 36\,000$ and $M_w/M_n = 1.05$ (MMA/BuA = 43/57), and the final step produced poly(MMA-blo-BuA-blo-MMA) with $M_n = 144\,000$ and $M_w/M_n = 1.09$ (MMA/BuA/MMA = 13/15/72). Thus, the MMA block was lengthened longer than expected, which may be attributed to partial deactivation of the living polymer chain end of poly(MMA/BuA). Thus, a mixture of MMA and BuA was added at once to the living polymer chain end of

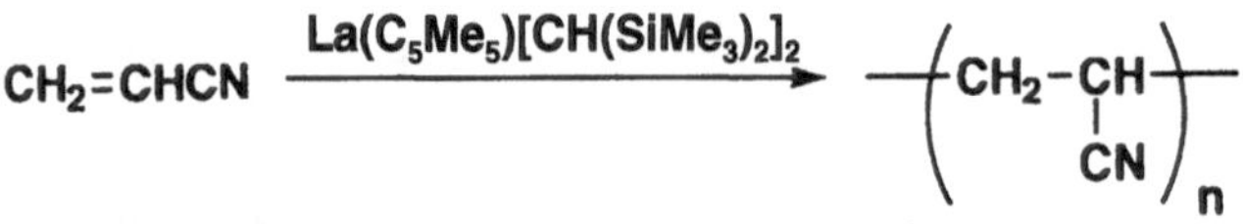

$$\text{CH}_2\text{=CHCN} \xrightarrow{\quad \text{La}(\text{C}_5\text{Me}_5)[\text{CH}(\text{SiMe}_3)_2]_2 \quad} \left(\text{CH}_2\text{--}\underset{\underset{\text{CN}}{|}}{\text{CH}} \right)_n$$

Scheme 7. Polymerization of acrylonitrile

poly(MMA) (BuA polymerized more rapidly than MMA), and it was found that this treatment allowed living triblock copolymerization to proceed preferentially and produced various triblock copolymers [35]. Table 6 shows the typical mechanical properties of the copolymers thus obtained. It is seen that homo-poly(MMA) has a large tensile modulus and great tensile strength, but is poor in elongation and Izod impact strength. Furthermore, homo-poly(MMA) shows no decrease in compression set (typical rubber shows ca. 20%). In contrast, the triblock copolymer (8 : 72 : 20) shows a 58% compression set, and its Izod impact strength is in the range "not break" (> 400 J/m), while its hardness is lowered to 20 JISA. Its most interesting property is increased elongation (163%). By contrast, for the triblock copolymers composed of MMA/BuA/MMA (6 : 91 : 3 and 25 : 51 : 24 ratios), the compression set increases to 97 or 101%, which indicates that these polymers are not elastic. The elongation of the 25 : 51 : 24 triblock copolymer is 81%, and that of the 6 : 91 : 3 copolymer is 246%. Thus a suitable composition ratio is required to obtain elastic triblock copolymers.

2.4 Polymerization of Acrylonitrile

Ren et al. [36] carried out polymerization of acrylonitrile using [(tBu-Cp)$_2$Nd CH$_3$]$_2$ to obtain an M_n of $2.6 - 8.2 \times 10^4$ at 10–28% yield, and showed that the product was atactic (rr = 26–30%, rm = 41–46%, mm = 27–30%) (Scheme 7). Hu et al. [37] found a bulky phenoxy complex, Sm [O-C$_6$H$_2$-2,6-(tBu)-4-Me]$_2$, initiates the atactic polymerization of acrylonitrile at 60 °C in toluene and obtained $M_n = 2.1 \times 10^4$ when the conversion reached 72%. According to recent work by Tanaka et al. [11], La(C$_5$Me$_5$)[CH(SiMe$_3$)$_2$]$_2$ has good catalytic activity for the polymerization of acrylonitrile (Scheme 6). However, the product was very poor in stereoregularity (rr = 26–29%, mr = 42–46%, mm = 25–32%), as was the case with such anion-type catalysts as BuLi and tBuMgCl [38].

3 Rare Earth Metal-Initiated Polymerizations of Alkyl Isocyanates

Polyisocyanates have attracted much attention owing to their liquid crystalline properties, stiff-chain solution characteristics, and induced optical activities associated with helical chain conformation (Scheme 8). Pattern and Novak [39] discovered that titanium complexes such as $TiCl_3(OCH_2CF_3)$ and $(C_5H_5)TiCl_2(OCH_2CF_3)$ initiate the living polymerization of isocyanates, producing polymers with narrow molecular weight distributions. When hexyl isocyanate was added to $TiCl_3(OCH_2CF_3)$, the polymerization took place at room temperature, with M_n increasing linearly with the initial monomer-to-initiator mole ratio or the monomer conversion ($M_w/M_n = 1.1-1.3$) over a wide range. Recently, Fukuwatari et al. [40] have found lanthanum isopropoxide to serve as a novel anionic initiator for the polymerization of hexyl isocyanate at low temperature ($-78\,°C$), which led to very high molecular weight ($M_n > 10^6$) and rather narrow molecular weight distribution ($M_w/M_n = 2.08-3.16$). Other lanthanide alkoxides such as $Sm(OiPr)_3$, $Yb(OiPr)_3$, and $Y(OiPr)_3$ also induced the polymerization of hexyl isocyanate. Furthermore, it was shown [40] that butyl, isobutyl, octyl, and *m*-tolyl isocyanates were polymerizable with lanthanum isopropoxide as the initiator. However, *t*-butyl and cyclohexyl isocyanates failed to polymerize with this initiator under the same conditions. When the reaction temperature was raised to ambient temperature, only cyclic trimers were produced at high yields. In an unpublished work, Tanaka et al. [11] showed that $La(C_5Me_5)_2[CH(SiMe_3)]_2$ also initiates the polymerization of butyl isocyanate and hexyl isocyanate at 50–60% yields.

It has been shown recently that the selective reductive homo-coupling polymerization of aromatic diisocyanates via one electron transfer promoted by samarium iodide in the presence of hexamethylphosphoramide $[PO(NMe_2)_3]$ (HMPA) can produce poly(oxamide)s in nearly quantitative yield (Scheme 9).

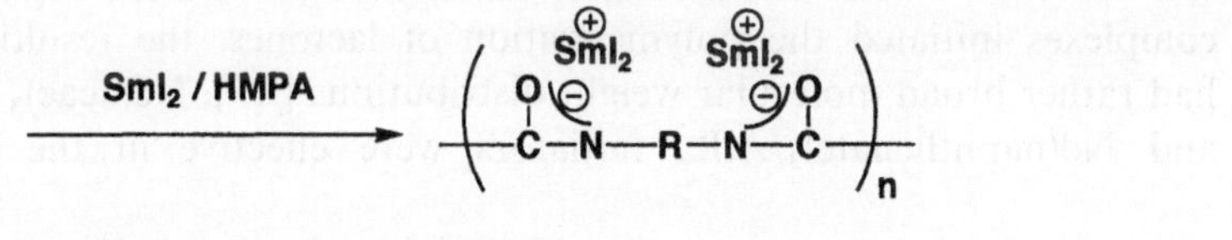

Scheme 8. Organolanthanide-initiated polymerization of alkylisocyanate

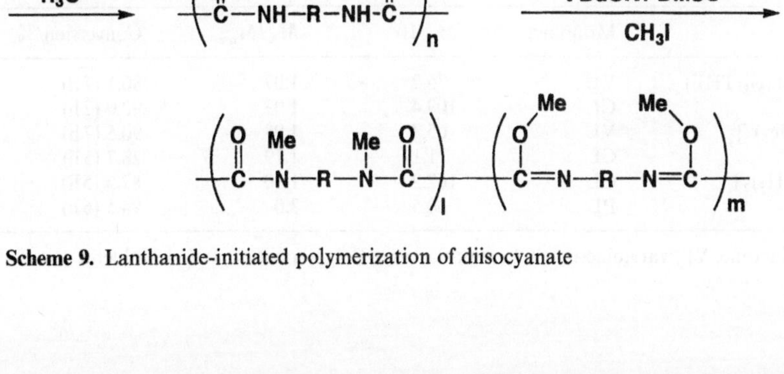

Scheme 9. Lanthanide-initiated polymerization of diisocyanate

The polymers obtained were insoluble in common organic solvents. The alkylation of poly(oxamide)s with methyl iodide or alkyl bromide at the presence of potassium *t*-butoxide gave readily soluble alkylated polymers at good yield. In either case, the alkylation was almost complete, and both N- and O-alkylation proceeded. The ratio of N- and O-methylation was found to be 64:36 by ^{1}H-NMR, and that of N- and O-alkylation was 3:1 by ^{13}C-NMR analysis. The SmI$_2$(HMPA)-initiated polymerization system was applied to a variety of diisocyanates including diphenylmethanediisocyanate, toluene 2,6-diisocyanate, 2,6-naphthyl diisocyanate, and *o*-toluidine diisocyanate, producing polymers with molecular weights of 2000–9000. TGA measurements showed T$_{d10}$ to be in the range of 248–320 °C [41, 42].

4 Ring-Opening Polymerization of Cyclic Monomers

4.1 Living Polymerization of Lactones

AlEt$_3$-H$_2$O or AlEt$_3$ catalyzed polymerization of β-methylpropiolactone and ε-caprolactone has been reported [43, 44], but this polymerization generally gives a broad molecular weight distribution. Yamashita et al. [45] explored the polymerization of various lactones [including β-propiolactone (PL), β-methyl-propiolactone (MePL), δ-valerolactone (VL), and ε-caprolactone (CL)] initiated by single organolanthanides obtained polymers with narrow molecular weight distributions, and found that VL and CL led to living polymerization yielding polymers with $M_w/M_n = 1.05$–1.10 at quantitative yields (Table 7). For ε-caprolactone, M_n obtained with the SmMe(C$_5$Me$_5$)$_2$(THF) or [SmH(C$_5$Me$_5$)$_2$]$_2$ system increased with increasing conversion, but M_w/M_n remained constant, irrespective of the conversion. For β-propiolactone, the use of YOR(C$_5$Me$_5$)$_2$ was more favorable. On the other hand, though divalent organolanthanide complexes initiated the polymerization of lactones, the resulting polymers had rather broad molecular weight distributions [46]. Nd(acac)$_3$·3H$_2$O/AlR$_3$ and Nd(naphthenate)$_3$/AlR$_3$ initiators were effective in the ring-opening

Table 7. Living polymerization of lactones with organolanthanide complexes

Initiator	Monomer	$M_n/10^3$	M_w/M_n	Conversion/%
SmMe(C$_5$Me$_5$)$_2$(THF)	VL	75.2	1.07	80.1 (7 h)
	CL	109.4	1.09	92.0 (7 h)
[SmH(C$_5$Me$_5$)$_2$]$_2$	VL	65.7	1.08	90.5 (7 h)
	CL	71.1	1.19	28.7 (5 h)
[YOMe(C$_5$H$_5$)$_2$]$_2$	CL	162.2	1.10	87.5 (5 h)
	PL	60.5	3.0	94.5 (5 h)

PL, propiolactone; VL, varelolactone; CL, caprolactone; Polymerization, 0 °C in toluene

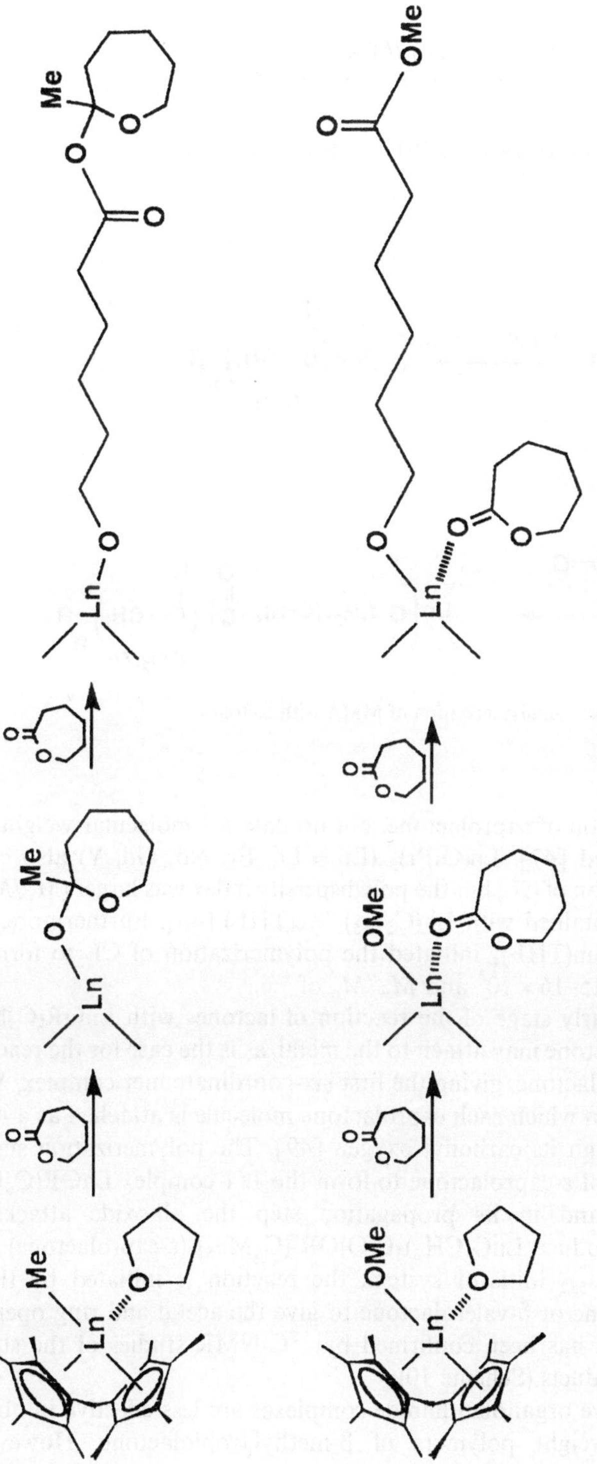

Scheme 10. Proposed mechanism for initiation of lactone polymerization

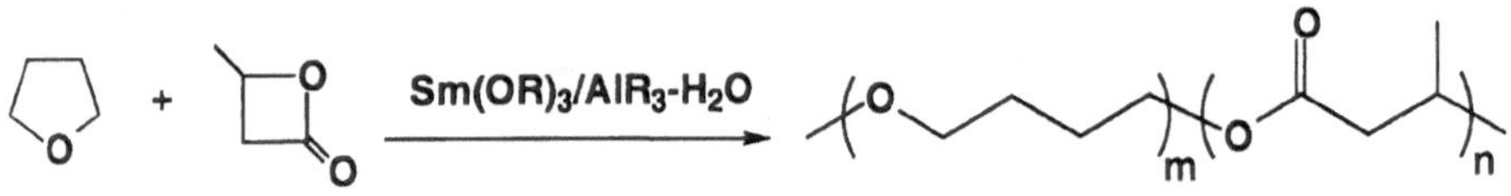

Scheme 11. Copolymerization of THF with β-methylpropiolactone

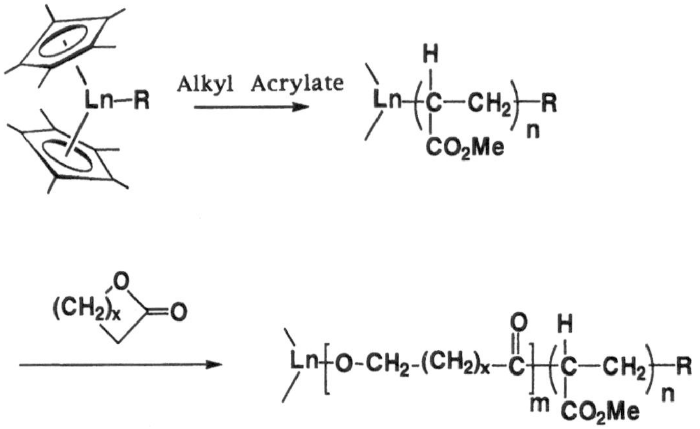

Scheme 12. Block copolymerization of MMA with lactones

polymerization of caprolactone, but no data for molecular weight distribution were reported [47]. Ln(OiPr)$_3$ (Ln = La, Pr, Nd, Gd, Y) also catalyzed the polymerization of CL, but the polydispersity index was larger (M_w/M_n = 1.2–2.4) than that obtained with Ln(C$_5$H$_5$)$_2$Me(THF) [48]. Furthermore, (2,6-tBu$_2$-4-Me-C$_6$H$_2$)$_2$Sm(THF)$_4$ initiated the polymerization of CL to form a polymer with M_n of 15–16 × 10^4 and M_w/M_n of 1.6.

At the early stage of the reaction of lactones with LnOR(C$_5$Me$_5$)$_2$, 1 mol of ε-caprolactone may attach to the metal, as is the case for the reaction of YCl$_3$ with ε-caprolactone, giving the first six-coordinate mer complex, YCl$_3$·(ε-caprolactone)$_3$ in which each caprolactone molecule is attached as a monodentate ligand through its carbonyl oxygen [49]. The polymerization starts with the attachment of ε-caprolactone to form the 1:1 complex LnOR(C$_5$Me$_5$)$_2$(ε-caprolactone), and in its propagation step the alkoxide attacks the C=O group to produce LnO(CH$_2$)$_5$C(O)OR-(C$_5$Me$_5$)$_2$(ε-caprolactone) [50]. In the SmMe(C$_5$Me$_5$)$_2$ initiator system, the reaction is initiated by the attack of ε-caprolactone or δ-valerolactone to give the acetal and ring opening follows. This process has been confirmed by ^{13}C-NMR studies of the stoichiometric reaction products (Scheme 10).

The above organolanthanide complexes are less effective in obtaining high molecular weight polymers of β-methylpropiolactone. However, random copolymerization of β-methylpropiolactone with THF occurred upon initiation with the $Sm(OiPr)_3/AlEt_3-H_2O$ system (1 : 5–1 : 10) (Scheme 11), although the $AlEt_3-H_2O$ system was completely inert in the polymerization of tetrahydrofuran (THF) [51]. The polymers produced were 43/57–24/76 in the lactone/THF ratio with an M_n of 10 000–200 000 and an M_w/M_n of 1.08–1.26. In this system, some cationic initiator is supposed to be formed when the two components are mixed.

Anionic block copolymerizations of MMA with lactones proceeded smoothly to give copolymers with M_w/M_n = 1.11–1.23 when the monomers were added in this order. However, when the order of addition was reversed, no copolymerization took place [3c], i.e., no addition of MMA to the polylactone active end group occurred (Scheme 12).

4.2 Polymerization of Tetrahydrofuran

The ring-opening polymerization of tetrahydrofuran (THF) was first reported by Meerwein [52], who used the trialkyloxonium salt as the initiator. Recently, Jin et al. [53] found that the $Ln(CF_3COO)_3/AlHBu_2$/propylene oxide (1/4/1) system allowed THF to polymerize at high yield (50–55%). Nomura et al. [54] succeeded in a quantitative one-pot transformation of the cationic growing centers of telechelic poly(THF) into anionic ones by using samarium (II) iodide/HMPA. The transformed poly(THF) macrodianion initiated the polymerization of t-butyl methacrylate (tBuMA) with quantitative initiation efficiency,

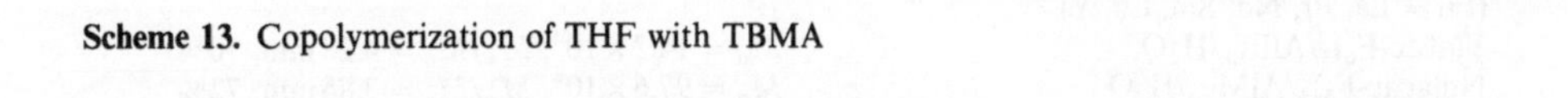

Scheme 13. Copolymerization of THF with TBMA

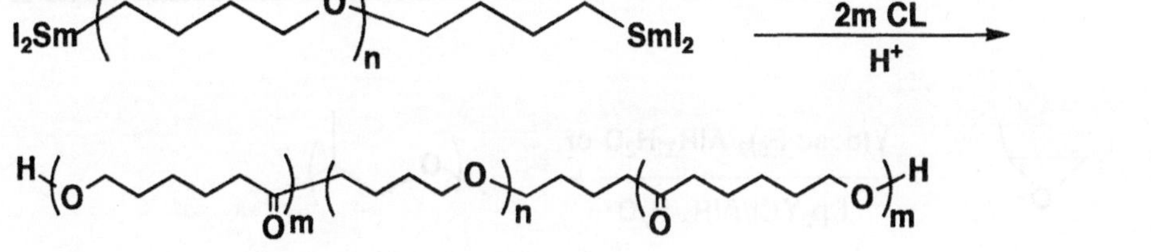

Scheme 14. Copolymerization of THF with caprolactone

and a unimodal triblock copolymer of *t*BuMA and THF was obtained (Scheme 13). According to Nomura et al. [55], the transformed poly(THF) macroanion is active in the living polymerization of CL producing an ABA-type triblock copolymer of CL with THF (Scheme 14).

4.3 Stereospecific Polymerization of Oxiranes

Such organolanthanide(III) complexes as $LnMe(C_5Me_5)_2$ or $LnH(C_5Me_5)_2$ do not initiate the polymerization of oxiranes, but more complex systems like $Ln(acac)_3/AlR_3/H_2O$ and $Ln(2\text{-}EP)_3/AlR_3/H_2O$ (2-EP; di-2-ethylhexylphosphate) do this with good initiation activity [56]. High molecular weight poly(ethylene oxide) is one of the common water-soluble polymers useful as adhesives, surfactants, plasticizers, and dispersants as well as for sizes. Poly(ethylene oxide) of $M_n = 2.85 \times 10^6$ was obtained with the $Y(2\text{-}EP)_3/AliBu_3/H_2O$ system at a ratio of $Y/Al/H_2O = 1/6/3$. The initiator activity depended upon the molar ratio of the components. Polymerization of propylene oxide was reported to proceed with the $Ln(acac)_3/AlEt_3/H_2O$ system, and it was found that light rare earth elements (Y, La, Pr, Nd, Sm) produced very high molecular weight poly(propylene oxide) at $Al/Ln = 6$ in a short period of time (2h) in toluene [57]. It was also found that the $Nd(acac\text{-}F_3)_3/AlR_3/H_2O$ (Ln = Y, Nd) systems [58] gave isotactic poly(propylene oxide), while the $Cp_2LnCl/AlR_3/H_2O$, $Sm(OiPr)_3/AlR_3/H_2O$, or $Y(2\text{-ethylhexanoate})_3/AlR_3/H_2O$ system produced relatively low molecular weight isotactic species of this polymer (Table 8, Scheme 15).

Random copolymerization of propylene oxide with ethylene oxide proceeded smoothly with the $Nd(2\text{-}EP)_3/AlEt_3/H_2O$ system at 80 °C [59]. From

Table 8. Polymerization of propylene oxide

Initiator	Polymer
$Ln(acac)_3/AlEt_3/H_2O$ (Ln = La, Pr, Nd, Sm, Eu, Y)	Head-to-tail, isotactic
$Y(acac\text{-}F_6)_3/AlEt_3/H_2O$	$M_n = 14.2 \times 10^4$, $M_w/M_n = 4.28$ mm, 70%
$Nd(acac\text{-}F_3)_3/AlMe_3/H_2O$	$M_n = 97.6 \times 10^4$, $M_w/M_n = 3.86$ mm, 72%
$Cp_2YCl/AlEt_3/H_2O$	$M_n = 24.8 \times 10^4$, $M_w/M_n = 4.69$ mm, 51%
$Y(2\text{-ethylhexanoate})_3/AlR_3/H_2O$	$M_n = 32.6 \times 10^4$, $M_w/M_n = 3.41$ mm, 44%
$Sm(OiPr)_3/AlEt_3/H_2O$	$M_n = 6.6 \times 10^4$, $M_w/M_n = 1.75$ mm, 36%

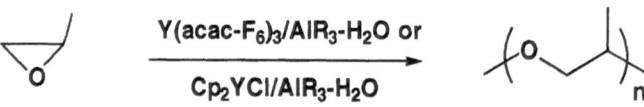

Scheme 15. Polymerization of propylene oxide

the copolymerization composition curve, the monomer reactivity ratios were evaluated to be $r_1(EO) = 1.60$ and $r_2(PO) = 0.45$. The conversion increased with an increase in the Al/Nd ratio and was saturated at a molar ratio of 16. The ^{13}C-NMR spectrum of the random copolymer clearly indicated the existence of the propylene oxide-ethylene oxide linkages in the polymer chains. Since the $Nd(2-EP)_3/AlEt_3/H_2O$ system generates a growing poly(propylene oxide) chain having a very long life, block copolymerization with ethylene oxide can be achieved successfully. The $Ln(acac)_3/AliBu_3/H_2O$ (1:8:4) systems, especially in the case of an Nd derivative, also initiated the polymerization of epichlorohydrin (EPH) to yield a polymer of $M_v = 16.5 \times 10^5$ with 21% crystallinity, and a remarkable solvent effect was observed [60]. Toluene is used preferably; aliphatic hydrocarbons are not suitable because poly(epichlorohydrin) precipitates from the solvent during the polymerization. The relative monomer reactivities evaluated for the propylene oxide (PO)-allyl glycidylether(AGE) system were $r_1(PO) = 2.0$ and $r_2(AGE) = 0.5$, and those for the epichlorohydrin-AGE system were $r_1(EPH) = 0.5$ and $r_2(AGE) = 0.4$ [60]. This combination of monomer reactivity ratios indicates that the polymerization with $Ln(acac)_3/AliBu_3/H_2O$ follows a coordination anionic mechanism, but that with the $AliBu_3/H_2O$ system follows a cationic polymerization mechanism. The ability to produce high molecular weight polymers is in the order Nd > La = Sm > Pr = Dy = Y > Gd = Yb > Eu.

Although the copolymerization of propylene oxide with CO_2 takes place effectively with organozinc additives or the (tetraphenyl) porphyrin-AlCl system [61], the copolymerization of epichlorohydrin with CO_2 seldom occurs with these catalysts. Shen et al. [62] showed that a rare earth metal catalyst such as the $Nd(2-EP)_3/AliBu_3$ (Al/Nd = 8) system was very effective for the copolymerization of epichlorohydrin with CO_2 (30–40 atm) at 60 °C (Scheme 16). The content of CO_2 in the copolymer reached 23–24 mol% when 1,4-dioxane was used as solvent.

The $ZnEt_2/H_2O$ [63], $AlEt_3/H_2O$ [64], and Cd salt [65] systems are well-known initiators for the polymerization of propylene sulfide. Shen et al. [66] examined this polymerization with the $Ln(2-EP)_3/AliBu_3/H_2O$ system and found that high molecular weight polymers were produced at a low concentration of Nd (6.04×10^{-3} mol/l) and at a ratio of $Nd/Al/H_2O = 1/8/4$. The polymerization activity decreased in the order Yb = La > Pr > Nd = Eu > Lu > Gd > Dy > Ho > Er. The ^{13}C-NMR spectrum indicated that β-cleavage occurs preferentially over the α-cleavage and the ratio of these ring openings changes little with the initiator system or the polymerization temperature. The polymers obtained were amorphous according to DSC and XRD analyses. Shen et al. [67] also showed that chlorotrimethylthiirane

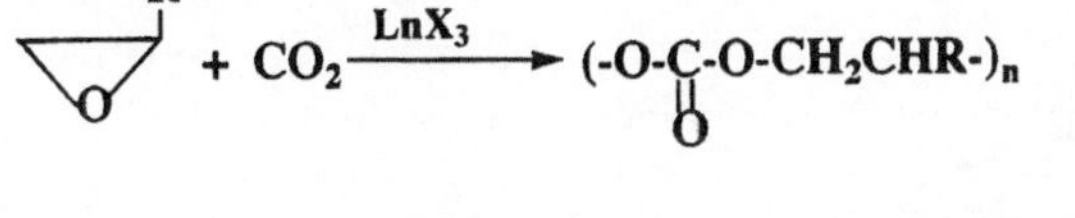

Scheme 16. Random copolymerization of oxirane with CO_2

(1-chloromethylethylene sulfide) can be polymerized to high molecular weight polymers ($[\eta] \times 10^2$ dl/g = 4–5) by using $Nd(P_{204})_3/AlR_3$ ($P_{204} = [CH_3(CH_2)_3$ $(CH_2CH_3)CHCH_2O]_2P(O)OH$) at a 1:12 ratio. Conversions were higher than 80%. Thus far, this monomer has not led to high molecular weight polymers regardless of catalyst.

4.4 Polymerization of Lactide

Shen et al. [68] was succeeded in the ring-opening polymerization of racemic D, L-lactide using $Nd(naphthenate)_3/AliBu_3/H_2O$ (1:5:2.5), $Nd(P_{204})_3/$ $AliBu_3/H_2O$, $Nd(P_{507})_3/AliBu_3/H_2O$ [$P_{507} = (iC_8H_{17}O)_2P(O)OH$], and Nd $(naphthenate)_3/AliBu_3/H_2O$ systems, obtaining polymers with molecular weights $M_n = 3.1–3.6 \times 10^4$ and conversions larger than 94%. When the $Ln(naphthenate)_3/AliBu_3/H_2O$ system was used, almost the same results were obtained irrespective of the metals used (La, Pr, Nd, Sm, Gd, Ho, Tm). Divalent $(2,6\text{-}tBu_2\text{-}4\text{-}Me\text{-}phenyl)_2Sm(THF)_4$ was also found to be active in the polymerization of D, L-lactide at 80 °C in toluene, giving M_n of $1.5–3.5 \times 10^4$ [69]. A more recent finding is that the $Ln(O\text{-}2,6\text{-}tBu_2\text{-}C_6H_3)_3/iPrOH$ (1:1–1:3) catalyst system initiates a smooth homo-polymerization of L-lactide, CL (caprolactone), and VL (valerolactone) and produces relatively high molecular weights ($M_n > 24 \times 10^3$) with low polydispersity indices ($M_w/M_n = 1.2–1.3$) (Scheme 17). Ring-opening polymerization of D, L-lactide was also carried out by using $Ln(OiPr_3)_3$ as the catalyst at 90 °C in toluene. The catalytic activity increased in the order La > Nd > Dy > Y and a molecular weight of 4.27×10^4 (conversion 80%) was obtained [70]. However, the molecular weight distribution is not clear at present. A kinetic study [10] showed that the rate of polymerization follows the first order in both monomer and initiator. The block copolymerization of CL with L-lactide proceeded effectively and produced a polymer with a very narrow molecular weight distribution ($M_w/M_n = 1.16$). On the other hand, the addition of CL to the living poly(L-lactide) end was not effective.

4.5 Polymerization of Cyclic Carbonates

Living polymerization of trimethylenecarbonate (TMC) and 2,2-dimethyltrimethylenecarbonate readily occurred in toluene at ambient temperature

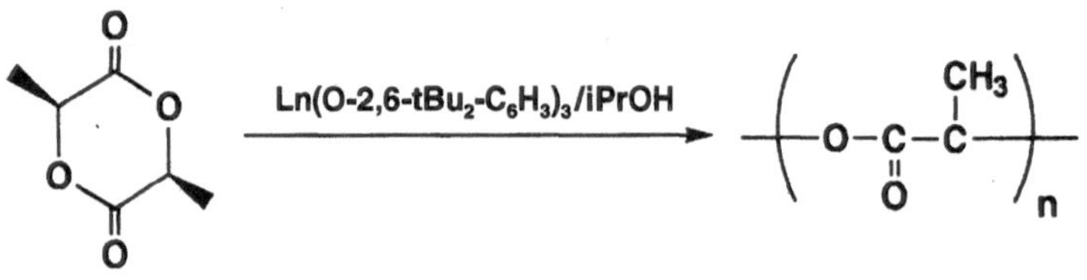

Scheme 17. Polymerization of L-lactide

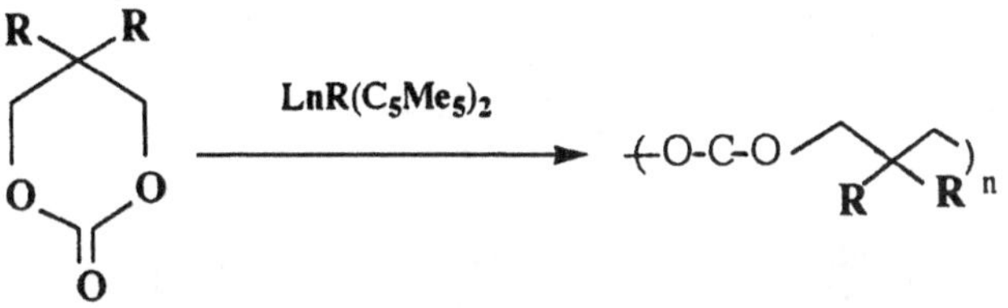

Scheme 18. Polymerization of cyclic carbonate

when $SmMe(C_5Me_5)_2(THF)$ or $YMe(C_5Me_5)_2(THF)$ was used as the initiator, yielding polymers with an M_n of $3-5 \times 10^4$ and very low polydispersity ($M_w/M_n = 1.0-1.1$) (Scheme 18) [71]. However, no detailed studies regarding its initiation and propagation mechanisms have been made as yet. Block copolymerization of trimethylenecarbonate with ε-caprolactone was carried out using $Ln(ethyl\ acetoacetate)_2(O\text{-isoPr})$ (Ln = Y, Nd) as the initiator [72]. CL was added to the poly(TMC) end or TMC to the living poly(CL) end, with the formed polymers having $M_n = 4-5 \times 10^4$ and $M_w/M_n = 1.2-1.4$.

5 Rare Earth Metal-Initiated Polymerization of Olefins, Dienes, and Acetylenes

5.1 Stereospecific Polymerization of Olefins

Bulky organolanthanide(III) complexes such as $LnH(C_5Me_5)_2$ (Ln = La, Nd) were found to catalyze with high efficiency the polymerization of ethylene [73]. These hydrides are, however, thermally unstable and cannot be isolated as crystals. Therefore, thermally more stable bulky organolanthanides were synthesized by introducing four trimethylsilyl groups into the Me_2Si-bridged Cp ligand, as shown in Fig. 8. The reaction of the dilithium salt of this ligand with anhydrous $SmCl_3$ produced a mixture of two stereo-isomeric complexes. The respective isomers were isolated by utilizing their different solubility in hexane; their structures were determined by X-ray crystallography. One isomer has a C_2 symmetric (racemic) structure in which two trimethylsilyl groups are located at the 2,4-position of the Cp rings, while the other has a C_1 symmetric structure in which two trimethylsilyl groups are located at the 2,4- and 2,3-positions of each Cp ring. Both were converted to alkyl derivatives by reaction with bis(trimethylsilyl) methyllithium [74] (Fig. 9). The Cp'-Sm-Cp' angle of the racemic-type precursor is 107°, which is about 15° smaller than that of non-bridged $SmMe(C_5Me_5)_2(THF)$.

Meso-type ligands were synthesized by forcing two trimethylsilyl groups to be located at the 3-position of the ligand with the introduction of two bridges (Fig. 10) [74]. Actually, the complexation of this ligand with YCl_3 yielded

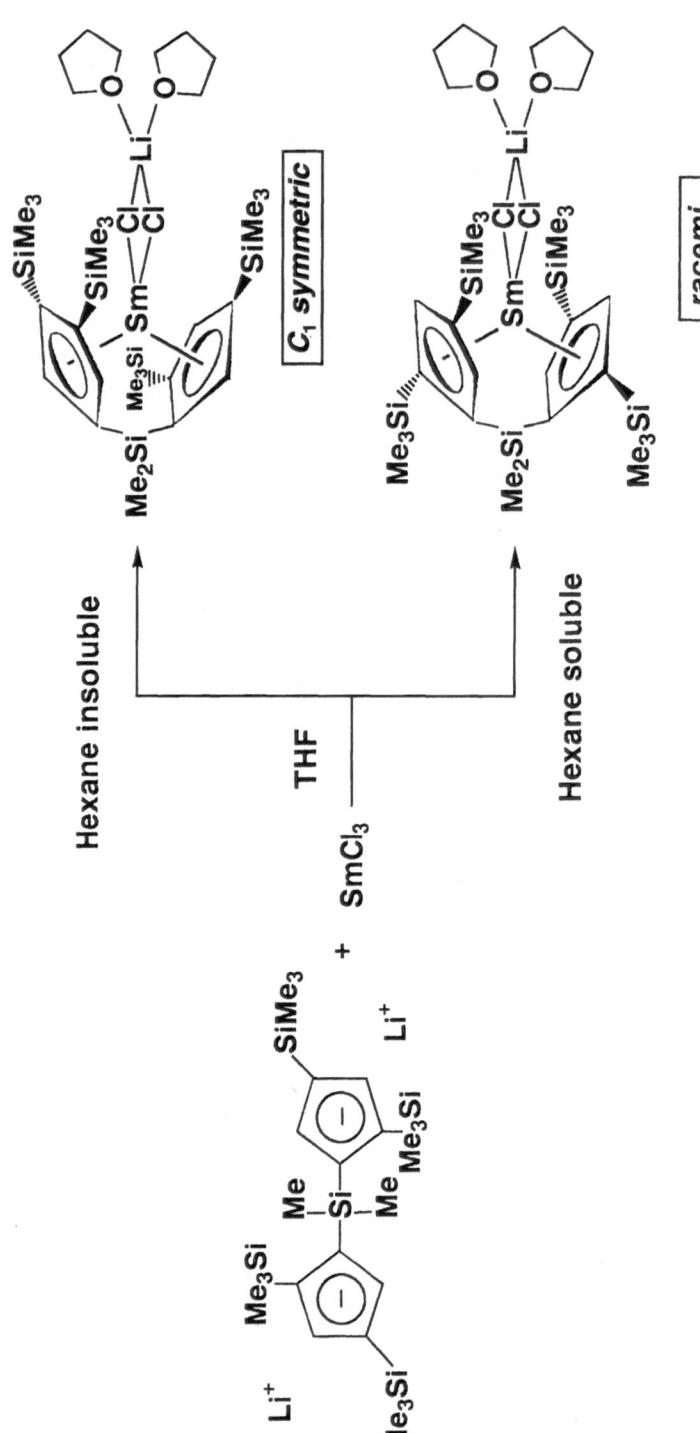

Fig. 8. Formation of organolanthanide(III) complexes

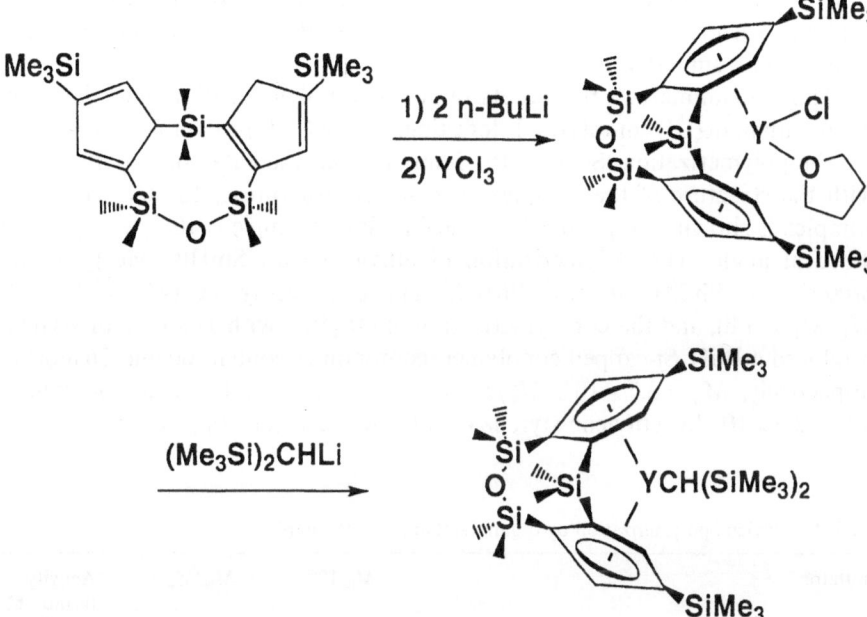

Fig. 9. Preparation of meso $Me_2Si(SiMe_2OSiMe_2)(3\text{-}SiMe_3\text{-}C_5H_2)_2YCH(SiMe_3)_2$

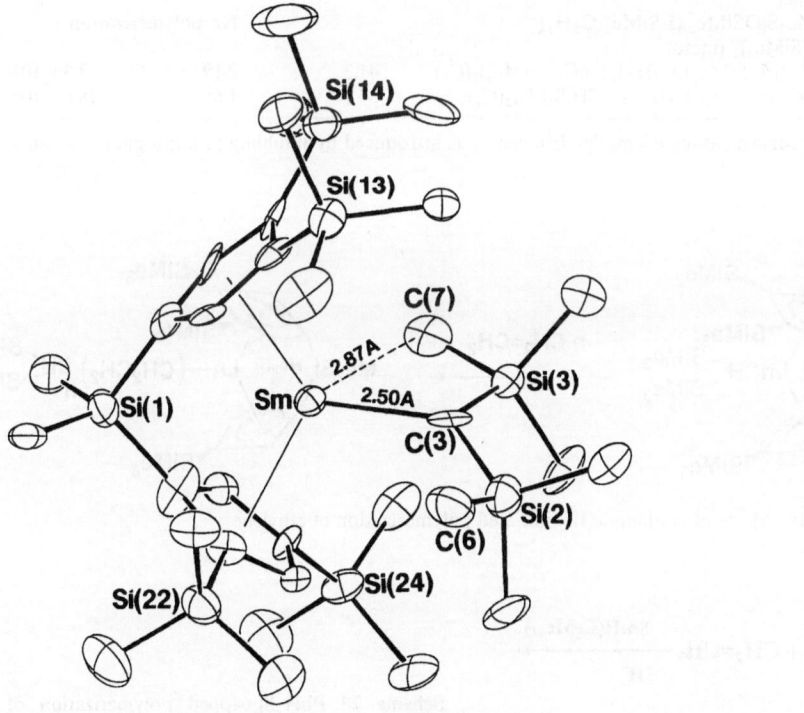

Fig. 10. X-ray structure of C_1 symmetric $Me_2Si[2(3),4\text{-}(SiMe_3)_2C_5H_2]_2SmCH(SiMe_3)_2$

a meso-type complex, and the structure of the complex was determined by X-ray analysis. Yasuda et al. [74] were also able to synthesize the meso-type alkyl complex in a similar way.

Table 9 summarizes the results of ethylene polymerization with these organolanthanide(III) complexes. Interestingly, only C_1-type complexes can initiate the polymerization (Scheme 19), implying that the catalytic activity varies with the structure of the complex. The X-ray structure of the C_1 symmetric complex is shown in Fig. 10, where the Cp′-Sm-Cp′ angle is 108°, a very small dihedral angle. The polymerization of ethylene with $SmH(C_5Me_5)_2$ in the presence of $PhSiH_3$ formed PhH_2Si-capped polyethylene ($M_n = 9.8 \times 10^4$, $M_w/M_n = 1.8$), and the copolymerization of ethylene with 1-hexene or styrene produced a PhH_2Si-capped copolymer (comonomer content 60 and 26 mol %, respectively; $M_n = 3.7 \times 10^3$, $M_w/M_n = 2.9$ for ethylene-1-hexene copolymer, $M_n = 3.3 \times 10^3$ for ethylene-styrene copolymer) (Scheme 20) [75].

Table 9. Ethylene polymerization by organolanthanide (III) complexes

Initiator	$M_n/10^4$	M_w/M_n	Activity (g/mol · h)
$(C_5Me_5)_2SmCH(SiMe_3)_2$		No polymerization	
$SiMe_2[2,4-(SiMe_3)_2C_5H_2]_2SmCH(SiMe_3)_2$ (racemic)		No polymerization	
$SiMe_2(Me_2SiOSiMe_2)(3-SiMe_3-C_5H_2)$ Y CH(SiMe_3)_2 (meso)		No polymerization	
$SiMe_2 [2(3),4-(SiMe_3)_2C_5H_2]_2SmCH(SiMe_3)_2(C_1)$	41.3	2.19	3.3×10^4
$SiMe_2[2(3),4(SiMe_3)_2C_5H_2]_2Y CH(SiM_3)_2(C_1)$	33.1	1.65	18.8×10^4

Initiator concentration, 0.2 mol%. Ethylene was introduced by bubbling at atmospheric pressure

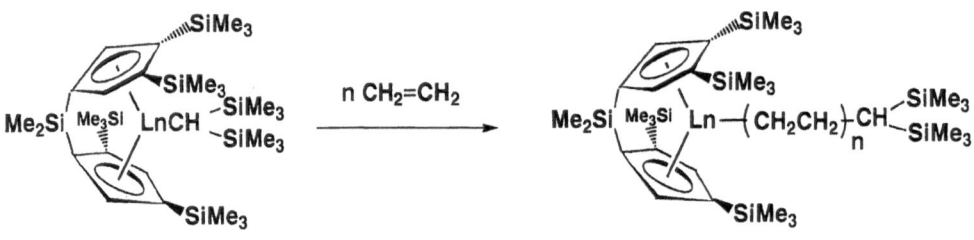

Scheme 19. Mode of lanthanide(III)-initiated polymerization of ethylene

$$PhSiH_3 + CH_2{=}CH_2 \xrightarrow[\text{H}^+]{SmH(C_5Me_5)_2}$$

$$PhSiH_2(CH_2CH_2)_nH$$

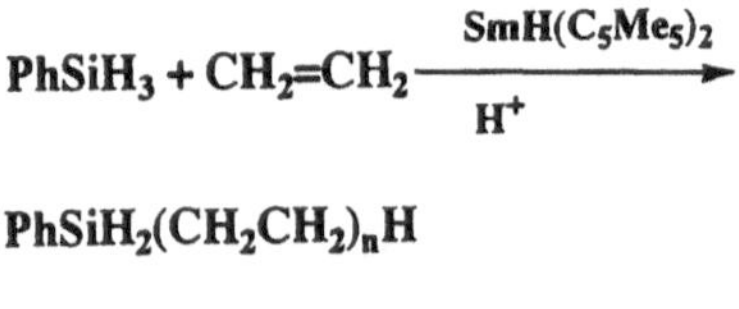

Scheme 20. PhH_2Si-capped polymerization of ethylene

Table 10 shows the results of polymerization of α-olefins catalyzed with trivalent complexes. When a bulkiler tBuMe$_2$Si group instead of the Me$_3$Si group was introduced into the yttrium complex, the racemic complex was formed exclusively [74c]. However, this alkyl complex did not react with olefins, and hence it was converted to a hydride complex by reaction with H$_2$. The complex obtained was reactive to various olefins and produced polymers at

Table 10. Polymerization of 1-pentene and 1-hexene

Monomer	Initiator	$M_n/10^3$	M_w/M_n
1-Pentene	Me$_2$Si(2-SiMe$_3$-4-tBu-C$_5$H$_2$)$_2$Sm(THF)$_2$	13	1.63
	Me$_2$Si [2(3),4-(SiMe$_3$)$_2$C$_5$H$_2$]$_2$Y CH(SiMe$_3$)$_2$	16	1.42
	Me$_2$Si(2-SiMe$_3$-4-tBu-C5H$_2$)$_2$YH]$_2$	20	1.99
1-Hexene	Me$_2$Si(2-SiMe$_3$-4-tBu-C$_5$H$_2$)$_2$Sm(THF)$_2$	19	1.58
	Me$_2$Si[(2(3),4-(SiMe$_3$)$_2$C$_5$H$_2$]$_2$Y CH(SiMe$_3$)$_2$	64	1.20
	Me$_2$Si(2-SiMe$_3$-4-tBu-C$_5$H$_2$)$_2$YH]$_2$	24	1.75

Initiator concentration, 0.2 mol% of monomer. Solvent, toluene; toluene/monomer = 5 (wt/wt)

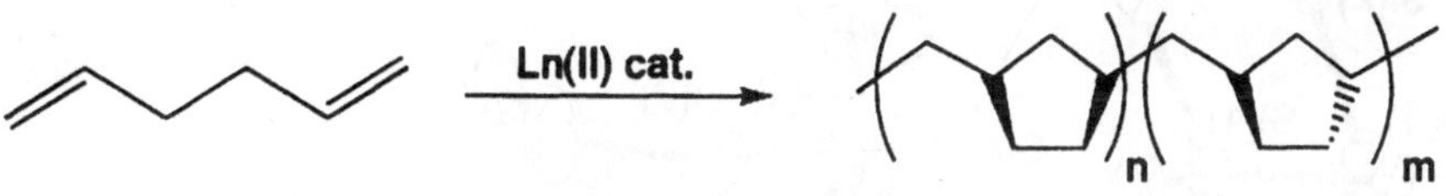

Scheme 21. Polymerization of 1,5-hexadiene

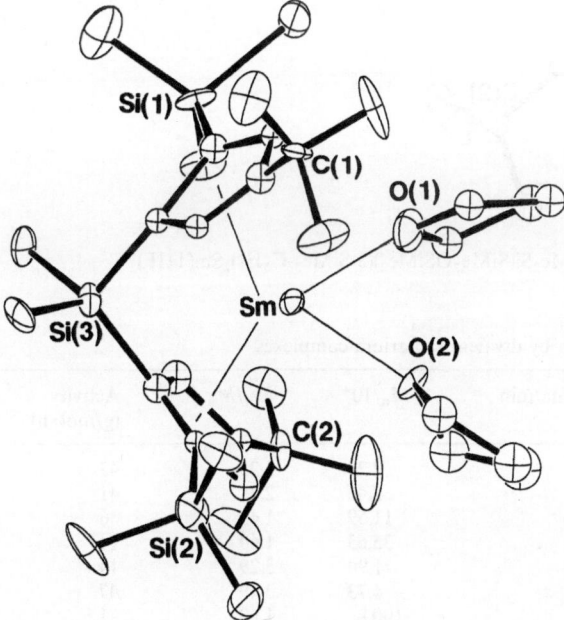

Fig. 11. X-ray structure of racemic Me$_2$Si(2-SiMe$_3$-4-tBu-C$_5$H$_2$)$_2$Sm(THF)$_2$

high yield. Poly(α-olefin) thus obtained was highly isotactic with an mmmm content >95%, when examined by ^{13}C-NMR at 34.91 ppm. It also initiated the polymerization of 1,5-hexadiene, yielding poly(methylene-1,3-cyclopentane) exclusively [76], whose M_n($M_n = 13.5 \times 10^4$) was much higher than that obtained with the Kaminsky-type catalyst ($M_n = 40\,000$) (Scheme 21). The conversions of poly(1-pentene) and poly(1-hexene) were much higher than those using rac-Me$_2$Si(2-SiMe$_3$-4-tBu-C$_5$H$_2$)$_2$YH [77] or [(C$_5$Me$_4$)SiMe$_2$-N-tBu)(PMe$_3$)ScH]$_2$ [78].

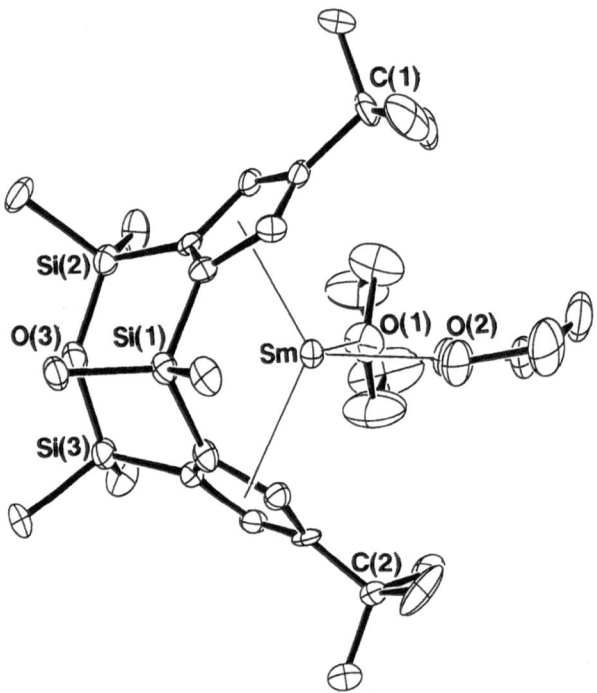

Fig. 12. X-ray structure of meso Me$_2$Si(SiMe$_2$OSiMe$_2$)(3-SiMe$_3$-C$_5$H$_2$)$_2$Sm(THF)$_2$

Table 11. Ethylene polymerization by divalent samarium complexes

Initiator	Time/min	$M_n/10^4$	M_w/M_n	Activity (g/mol·h)
(C$_5$Me$_5$)$_2$Sm(THF)$_2$	1	2.28	1.25	43
	3	2.46	2.28	41
Racemic	1	11.59	1.43	6
	3	35.63	1.60	14
Meso	5	1.94	3.29	14
	10	4.73	3.49	47
C$_1$ symmetry	15	100.8	1.60	1.6
	30	> 110.0	1.64	1.1

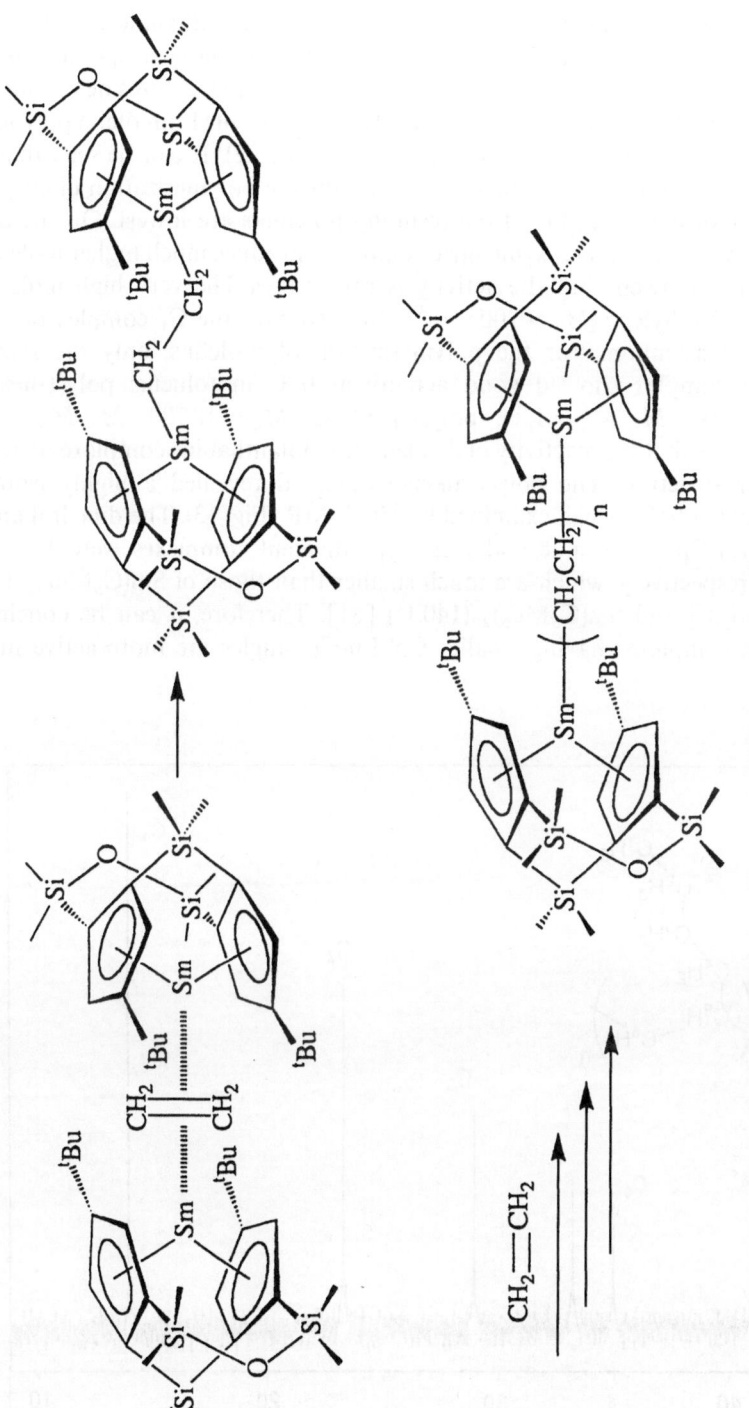

Scheme 22. Mode of lanthanide(II)-initiated polymerization of ethylene

Racemic, meso, and C_1 symmetric divalent organolanthanide complexes can also be synthesized by allowing the dipotassium salt of the corresponding ligand to react with SmI_2 [79]. Figures 11 and 12 show their structures determined by ^1H-NMR and X-ray analyses. Table 11 shows the results from olefin polymerization with divalent samarium complexes (Scheme 22). It can be seen that the meso-type complex has the highest activity for the polymerization of ethylene, but the molecular weights of the resulting polymers are lowest. On the other hand, the racemic and C_1 symmetric complexes produce much higher molecular weight polyethylene but the activity is rather low. The very high molecular weight polyethylene ($M_n > 100 \times 10^4$) obtained with the C_1 complex deserves particular attention. For the polymerization of α-olefins, only the racemic divalent complex showed good activity at $0\,°C$ in toluene: poly(1-hexene) $M_n = 24\,600$, $M_w/M_n = 1.85$; poly(1-pentene) $M_n = 18\,700$, $M_w/M_n = 1.58$. Thus, we see that the reactivity of divalent organolanthanide complexes depends on their structure. The poly(1-alkene) obtained revealed a highly isotactic structure (> 95%) when examined by ^{13}C-NMR (Fig. 13). The dihedral angles of Cp'-Ln-Cp' of racemic and meso-type divalent complexes were $117°$ and $116.7°$, respectively, which are much smaller than those of $Sm(C_5Me_5)_2(THF)$ ($136.7°$) [80] and $Sm(C_5Me_5)_2$ ($140.1°$) [81]. Therefore, it can be concluded that the complexes having smaller Cp'-Ln-Cp' angles are more active in the

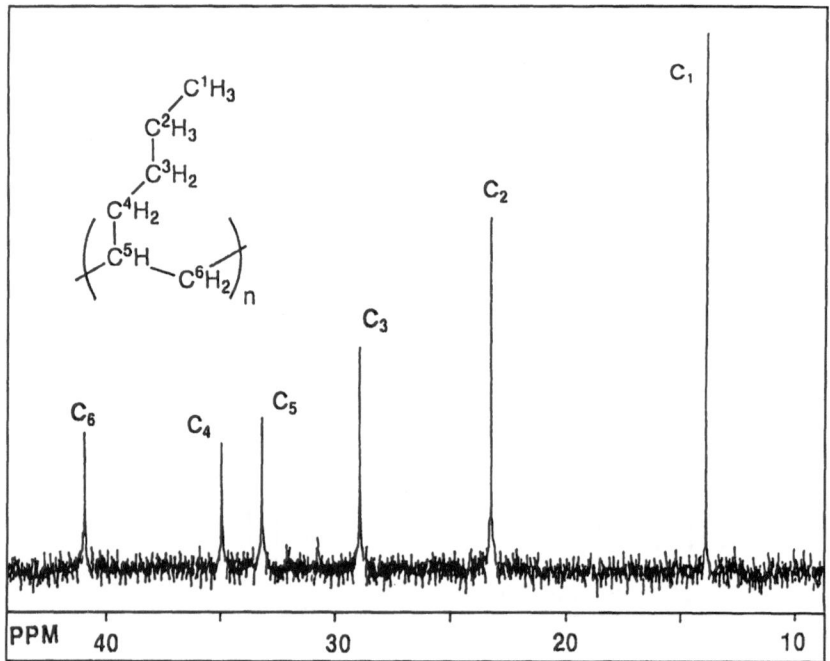

Fig. 13. ^{13}C-NMR spectrum of isotactic poly(1-hexene)

polymerization of ethylene and α-olefins. 1,5-Hexadiene was polymerized smoothly by the catalytic action of rac-$Me_2Si(2-SiMe_3-4-tBu-C_5H_2)_2Sm(THF)_2$ to produce poly(methylene-1,3-cyclopentane) at a ratio of $cis/trans = 55/45$ [74c].

5.2 Polymerization of Styrene

Styrene polymerization was performed by using binary initiator systems such as $Nd(acac)_3$-AlR_3 or $Nd(P_{507})_3$-AlR_3; syndio-rich polystyrene was obtained at a ratio of $Al/Nd = 10$–12 (Scheme 23) [82]. More recently, it has been shown that the $Gd(OCOR)_3/iBu_3Al/Et_2AlCl$ catalytic system initiates the copolymerization of styrene with butadiene, but produces only atactic polystyrene [83]. The $Sm(OiPr)_3/AlR_3$ or $Sm(OiPr)_3/AlR_2Cl$ ($Sm/Al = 1$–15) catalytic system also initiates the polymerization of styrene, producing a high molecular weight polymer ($M_n = 300\,000$), low in polydispersity but atactic in stereoregularity [84]. The cationic polymerization of styrene using $Ln(CH_3CN)_9(AlCl_4)_3$ (CH_3CN) was also examined [85]. It was found that the activity increased in the order La (conversion 73%) > Tb = Ho > Pr = Gd > Nd > Sm = Yb > Eu (conversion 54%), while M_n decreased when the polymerization temperature increased from 0 (20×10^3) to 60 °C (13×10^3). A more recent study [86] has shown that the single-component catalyst $[(tBuCp)_2LnCH_3]_2$ (Ln = Pr, Nd, Gd) initiated the polymerization of styrene at a relatively high temperature (70 °C), with a conversion of 96% for $[(tBuCp)_2NdCH_3]_2$ and M_n of 3.3×10^4 [86], though stereoregularity was very poor. The activity varied greatly with the lanthanide element, and the catalytic activity increased in the order Nd > Pr > Gd ≫ Sm, Y (the Sm and Y complexes showed practically no activity). Therefore, the reaction is considered to follow the radical initiation mechanism. Styrene polymerization was also performed successfully using the single-component initiators, $[(Me_3Si)_2N]_2Sm(THF)_2$, $[(Me_3Si)_2CH]_3SM$, and $La(C_5Me_5)$-$[CH(SiMe_3)_2]_2(THF)$ at 50 °C in toluene with no addition of cocatalyst. The resulting polymers had an $M_n = 1.5 - 1.8 \times 10^4$ with $M_w/M_n = 1.5 - 1.8$, and were only atactic [37, 80]. Thus no success has yet been achieved in synthesizing syndiotactic polystyrene with rare earth metal complexes, in contrast to the synthesis of highly syndiotactic polystyrene with $(C_5Me_5)TiCl_3/(AlMe-O-)_n$ system (syndiotacticity > 95%) [87, 88].

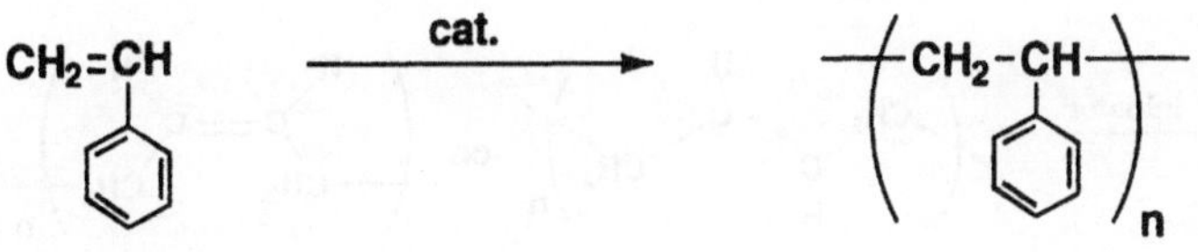

Scheme 23. Polymerization of styrene

5.3 Stereospecific Polymerization in Conjugate Dienes

Organolanthanide(III)-based binary initiator systems were used by Yu et al. [89] for stereospecific polymerization of butadiene (Table 12) and isoprene. Typically, the polymerization of butadiene catalyzed by $C_5H_5LnCl_2 \cdot THF/ AlR_3$ yielded polymers with a *cis*-1,4-content as high as 98% (Scheme 24). The polymerization activity decreased in the order Nd > Pr > Y > Ce > Gd and $iBu_2AlH > iBu_3Al > Et_3Al > Me_3Al$, while the viscosity of the polymer decreased in the order $Et_3Al > iBu_3Al > iBu_2AlH$. Although the $NdCl_3/iBu_3Al$ system exhibited practically no initiating activity, the use of $NdCl_3/PrOH$ instead of the solvent-free metal chloride brought about high polymerization activity and high stereoregularity in the *cis*-1,4-polymerization of butadiene [90]. The M_n of the polymer increased linearly with increasing conversion and reached 1530×10^3 at 85% conversion when $NdCl_3 \cdot iPrOH/AlEt_3$ (1:10) was used in heptane at $-70\,°C$, but M_w/M_n (1.8–2.5) showed no change with conversion. The number N of polymer chains per metal atom was 1.09–1.43 at $-70\,°C$, and increased to 2.0–3.0 when the polymerization temperature was raised to $0\,°C$. Most noteworthy is a very high *cis*-content realized at $-70\,°C$, which amounted to 99.4%. This indicates the existence of the anti-π-allyl-Nd species rather than the syn-π-allyl-Nd species in the polymerization system. The *cis*-1,4-content of poly(butadiene) increased as the $AlEt_3$ concentration was lowered [90].

The arene organolanthanide system, $Nd(C_6H_6)(AlCl_4)_3/AliBu_3$ (Al/Nd = 30), also induces the catalytic polymerization of isoprene to give *cis*-1,4-polymers having 92–93% selectivity at low conversion (17–36%). Neither the

Table 12. Polymerization of butadiene

Initiator	Polymer
$C_5H_5LnCl_3/AlR_3$ (Ln = Ce, Pr, Nd, Gd, Y; R = iBu, Et)	*cis*-1,4 > 98%
$NdCl_3/iPrOH/AlEt_3$	*cis*-1,4, 99.4% $M_n = 1.5 \times 10^5$, $M_w/M_n = 1.0 - 2.7$
$(iPrO)_3$ Gd complex,	*cis*-1,4, 92.5%
Dy complex,	*cis*-1,4, 84.6%
Er complex,	*cis*-1,4, 82.8%
$(CF_3COO)_2LnCl/EtOH/AlEt_3$ Nd Complex, (Ln = La, Ce, Pr, Nd, Sm, Eu)	*cis*-1,4, 97.5%
$Nd(octanoate)_3/AlEt_2Cl/AlEt_3$	*cis*-1,4 > 98%
$[(CF_3)_3(\mu\text{-}O)_3(\mu_3\text{-}O)_3NdAlEt_2 \cdot 2THF]_2$	*trans*-1,4, 79.2%
Didymium(versate)$_3$/MgBu$_2$	$M_n = 5 \times 10^4$–8×10^4 $M_w/M_n = 1.5$, *trans*-1,4, 96.9%
$LnCl_3$ or $Ce(acac)_3/AlR_3$	*trans*-1,4, 65–70%

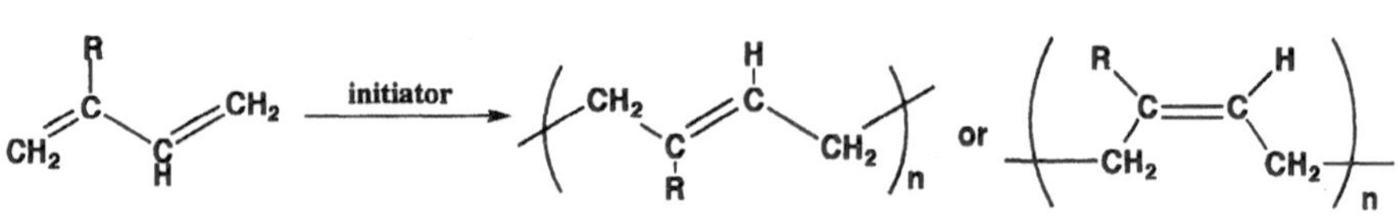

Scheme 24. Organolanthanide-initiated polymerization of butadiene derivatives

$Nd(C_6Me_6)(AlCl_4)_3/AliBu_3$ (1:30) nor $NdCl_3/AlCl_3/AliBu_3$ (1:3:30) system showed catalytic activity in the polymerization of isoprene [91]. The random copolymerization of isoprene with butadiene went smoothly using the $Nd(C_6H_6)(AlCl_4)_3/AliBu_3$ system and produced an isoprene/butadiene(1:4) copolymer at high yield, but, no data for M_w/M_n and M_n were reported [92]. The $(\beta\text{-}CH_3\text{-}\pi\text{-allyl})_2LnCl_5Mg_2(TMEDA)_2/AlR_3$ and $(\text{allyl})_4Li$ systems also initiate the 1,4-polymerization of isoprene in 50% stereoregularity at high conversion [92]. Highly selective cis-1,4-polymerizations of conjugated dienes were obtained using the homogeneous $(CF_3COO)_2NdCl \cdot EtOH/AlEt_3$ (1:7) initiator system, i.e., 97.5% cis-selectivity for butadiene and 96.7% for isoprene. Although bimetallic species like $(CF_3COO)EtNd(\mu\text{-}Cl)(\mu\text{-}H)AlEt_2$ are assumed to be active, their exact structure is still unknown [93]. The molecular structure of dimer complexes, $[(CF_3C)(\mu_2\text{-}O)_2(\mu_3\text{-}O)_4YAlEt_2(2THF)]_2$ and $[(CF_3)(\mu_2\text{-}O)_3(\mu_3\text{-}O)_3NdAlEt_2(2THF)]_2$, generated during the reaction between $(CF_3COO)_2NdCl$ and $AlEt_3$, has recently been elucidated by X-ray analysis. However, these complexes are considered to be by-products because they produce polymers having low stereoregularity at low yield [94].

A remarkable solvent effect was observed regarding the activity of the $Nd(OCOC_7H_{15})_3/Et_2AlCl/iBu_3Al$ system, which initiates the cis-1,4-polymerization of butadiene (> 98%) and isoprene (> 95%). Aliphatic compounds such as pentane, 1-pentene, and 2-pentene act as good solvents, while aromatic compounds such as toluene and mesitylene act as inhibitors [95]. The aromatic compounds attached to the metal center may be responsible for the remarkable suppression of polymerization, as was the case in polymerizations using cobalt catalysts [96]. A binary initiator system, $LnCl_3[(BuO)_3PO]_3/AliBu_3$, and a ternary system, $Ln(\text{naphthenate})_3/AliBu_3/Al_2Et_3Cl_3$ (naphthenate; $C_{10}H_7COO$), also act as good initiators for the cis-1,4-polymerization of isoprene [97]. In both cases, the polymerization activity varies with the nature of the metal in the order Nd > Pr > Ce > Gd > La ≫ Sm > Eu. Thus, the activity increases with increasing electronegativity and does so with decreasing M^{3+} ionic radius, except for Sm^{3+} and Eu^{3+} which can be easily converted to the M^{2+} species in the presence of a reducing agent. No relationship was observed between the initiating activity and the Ln–O or Ln–Cl bond energy determined by IR and laser Raman spectroscopy.

Block copolymerization of butadiene with isoprene (32:68–67:33), producing high cis-1,4-polymers, was also successfully carried out with the $Ln(\text{naphthenate})_3/AliBu_3/Al_2Et_3Cl_3$ system at temperatures of $- 78$ to $33\,°C$. Noteworthy is the relatively long lifetime of this initiator. Thus, it was possible to copolymerize isoprene 1752 h after the polymerization of butadiene. The $(iPrO)_2HLn_2Cl_3HAlEt_2$ species (Ln = Gd, Dy, Er, Tm) [98] prepared from either $Ln(iPrO)_3/Et_2AlCl/Et_3Al$ or $(iPrO)_2LnCl/Et_3Al$ can also initiate the cis-1,4-polymerization of butadiene and isoprene. The most probable structure of this complex as evidenced by X-ray analysis is $(iPrO)HLnEt(Cl)AlHEt(Cl)LnCl(OiPr)$. The polymerization activity decreased in the order Gd > Dy > Er > Tm and the cis-content ranged from 92 to 95% in the case of Gd derivatives. Random copolymerization of butadiene with isoprene was also performed using $Nd(C_6H_6)(AlCl_4)_3/AlR_3(Al/Nd = 30)$ in benzene. Both monomers were incorporated in the copolymer selectivity with cis-1,4-butadiene 96.1–96.4% and cis-1,4-isoprene 97.5–98.3%. The conversion increased with an increase in the polymerization temperature from 0 (10%) to 80 °C (80–100%) [99].

Some rare earth metal-based initiators induce the trans-1,4-polymerization of conjugated dienes at high yield. The $Ce(acac)_3/AlEt_2Cl$ system as well as the $CeCl_3/AlEt_2Cl$ and $GdCl_3/AlEt_3$ systems were most effective in the polymerization of isoprene with high selectivity (91–97%) [100]. The marked difference in the selectivity between the $Ce(acac)_3/AlEt_3$ and the $C_5H_5LnCl_2/AlR_3$ or $(iPrO)_2LnCl/AlR_3$ initiator systems may be due to a specific action of small amounts of water present in the system. Actually, metal compound hydrates like $NdCl_3 \cdot 6H_2O$, $PrCl_3 \cdot 6H_2O$, and $UO_2(C_2H_3O_2)_2 \cdot 6H_2O$ can initiate trans-polymerization in the presence of AlR_3 [101]. Prolonged aging of the initiator system decreased the activity significantly, presumably owing to an irreversible self-reaction of the intermediate generated from the organolanthanide and water. A recent study [101] on the effect of water has revealed that maximum conversion was attainable at an $H_2O/AlEt_2Cl$ ratio of 1.1–1.2, which produces $(AlEt-O-)_n(AlCl-O)_m$ species. However, the molecular weight was independent of the amount of water added to $AlEt_2Cl$.

It was found that didymium versatate (didymium refers to a rare earth mixture containing 72% Nd, 20% La, and 8% Pr)/MgR_2 (Mg/rare earth metal = 1/0.1 mol/mol) mixtures initiated the trans-1,4-polymerization of butadiene (> 97%) to produce polymers of $M_n = 50000$–8000 with $M_w/M_n = 1.5$ [102]. As an extension of this study, the butadiene/styrene block copolymerization was successfully carried out using a didymium versatate/MgR_2/BuLi at a ratio of 0.06:1:1.2 [102]. Furthermore, block copolymers of cis/trans-butadiene were synthesized by using an initiator system which was prepared in two steps, i.e., first a didymium versatate was added to $MgBu_2$ at a 0.1:1.0 ratio and then 1.5 mol equivalents of $Al_2Et_3Cl_3$ were added to the mixture in hexane after 60 min. Julemont et al. [103] obtained stereoblock polymers containing cis- and trans-blocks using nickel trifluoroacetate initiators.

Polymer-supported lanthanide initiators have been used frequently in the polymerization of conjugated dienes. It is well known that the efficiency of the activity initiating center is influenced by the nature and valence state of the transition metal, i.e., the type of ligands attached to the transition metal and the type of organometallic compounds. Manipulating these parameters effectively led to successful developments of more highly active initiator systems. In contrast to low molecular weight metal initiator systems, the environment of the metal ions in the active centers of polymer-metal initiator systems is influenced by the polymer ligand. Typical systems are styrene-divinylbenzene and styrene-acrylic acid copolymers. Neodium- and praseodymium-supported polystyrene showed high activity in the polymerization of butadiene and yielded cis-1,4-polybutadiene (97%) in the presence of $AliBu_3$ and $AlEt_2Cl$ [104]. Their polymerization activity is higher than that of a typical homogeneous low molecular weight initiator system, $Nd(acac)_3/AlR_3$. The polymerization of butadiene using the Nd compound supported by poly(styrene/acrylic acid) (3 : 1) in the presence of AlR_3 (R = Et, iBu) (Al/Nd = 200 mol/mol) and alkyl chloride ($PhCH_2Cl$) produced the cis-1,4-polymers with 98% selectivity at > 82% conversion when the ratio of the additives was 1 : 200 : 3.5 [104]. The IR spectra of the reaction products of poly(styrene/acrylic acid) and poly(ethylene/acrylic acid) with $NdCl_3$ showed new bands at 1550 and $1540 \, cm^{-1}$ as well as 1426 and 1420 cm, respectively. These bands are assignable to antisymmetric and symmetric stretching vibrations of COO^- groups due to the formation of the complexes. Since the increase in the covalency of metal-oxygen bonds always accompanies an increase in v_s (COO^-) and a decrease in v_{as} (COO^-), the polymer-supported neodymium complexes are considered to have a bidentate or tridentate carboxylate structure.

It was shown that the activity of poly(styrene/acrylic acid)-supported Nd/alkylaluminum halides/$AliBu_3$ depends largely on the kind of alkylaluminum halides and increases in the order $AlEt_2Cl > AlEtCl_2 > AlEt_2Br$ [105]. The effect of alkyl halides on the activity decreased in the order $Ph_3Cl > PhCH_2Cl > CH_2 = CHCH_2Cl \gg CCl_4 \gg (CH_3)_2CHCCl = C(CH_3)C$ H_2CH_2Cl. Thus, cis-1,4-polybutadiene with more than 98% selectivity was made available [106]. The poly(styrene/acrylic acid)-supported Nd complex also effectively initiated the polymerization of isoprene in the presence of Ph_3Cl and $AliBu_3$ at a ratio of 1 : 3.5 : 100 and yielded a polymer of high cis-1,4-content (96.4%) [106]. The polymerization efficiency of the poly [styrene/(2-methyl sulfinyl) ethyl-methacrylate] (S = O content, 7.9–9.7%)-supported $NdCl_3$ (Nd/S = O = 0.15–0.80(mol/mol) in the presence of $AliBu_3$(Al/Nd = 100–250) is two or three times higher than that of the $NdCl_3(DMSO)_4/AliBu_3$ system for the polymerization of butadiene. The cis-1,4-content reaches 98.1%. It is estimated for this reaction that 0.23 mol $NdCl_3$/mol of the functional group is bound as S = O–$NdCl_3$. When the content of (2-methylsulfinyl) ethyl methacrylate reached 34% in the copolymer, the butadiene conversion decreased markedly [107]. Neodydium complexes attached to carboxylated polystyrene cross-linked with divinylbenzene are known to be active in the presence of $AliBu_3/AlEt_2Cl$ (20 : 1 ratio) for the cis-1,4-polymerization (97%) of butadiene [108]. The contents of $Nd(O_2C$-$PS)_3$ and $Ce(O_2C$-$PS)_3$ units in the resin were 0.86 and 1.19 mmol of Nd and Ce per gram of polymer, respectively. After the completion of polymerization, dry benzene was added to dilute the polybutadiene and then EtOH was added to cleave the metal-polybutadiene bond. The recovered polymer support is recyclable and can be used four times. The cis-selectivity is maintained during these cycles without significant loss of activity. Neodymium derivatives of carboxylated polyethylene obtained from commercially available oxidized polyethylene also produced recyclable polymers when the ratio of $AliBu_3/AlEt_2Cl/Nd$ was fixed at 40 : 20 : 1 [108].

Butadiene-styrene copolymerization was attempted using the L_3Ln-RX-AlR_3 system [109]. Especially, $(CF_3COO)_3Nd/C_5H_{11}Br/AlisoBu_3$ (1 : 3 : 15) was found to be active in this type of copolymerization, with the cis-content of butadiene amounting to 97.8% and the styrene content to ca. 32%. However, for the isoprene/styrene system, the *trans*-1,4-polyisoprene copolymer was produced exclusively.

5.4 Stereospecific Polymerization of Acetylene and Its Derivatives

Polyacetylene (PA) is one of the simplest conjugated polymers, useful for manufacturing lightweight high-energy density plastics for storage batteries, solar energy cells, etc. Acetylene can be polymerized to produce high-*cis* PA film with $Ti(OiBu)_4/AlEt_3$ [110] or $Co(NO_3)_2/NaBH_4$ [111] at temperatures lower than $-78\,°C$. Recently, it has been reported that $Ln(naphthenate)_3/AlR_3/$ Donor (1:10:2-3) (Donor = acetone, ether, ethyl acetate) systems can also initiate stereoregular *cis*-polymerization of acetylene at $30\,°C$, which leads to silvery metallic film (Scheme 25, Table 13) [112,113]. The polymer yield increased with increasing polymerization temperature (-15 to $45\,°C$). A *cis*-polyacetylene with 95% selectivity was obtained when the Al/Ln ratio was adjusted to ca. 5. The polymerization activity decreased in the order Y = Ce > Nd = Tb > Pr > La > Lu > Gd > Tm = Er > Ho = Yb = Eu > Sm > Dy. The *trans*-content of the film amounted to 100% when the temperature was raised to $180\,°C$. The elements leading to PA film with a *cis*-content exceeding 95% are La, Pr, Nd, Sm, Gd, Tb, Dy, Ho, Er, Tm, and Y. The electrical conductivity of the film was $294 \times 10^{-8}\,S\,cm^{-1}$ for La, $181 \times 10^{-8}\,S\,cm^{-1}$ for Nd, $194 \times 10^{-8}\,S\,cm^{-1}$ for Gd, $490 \times 10^{-8}\,S\,cm^{-1}$ for Tb, and $184 \times 10^{-8}\,S\,cm^{-1}$ for Tm. Differential scanning calorimetry revealed

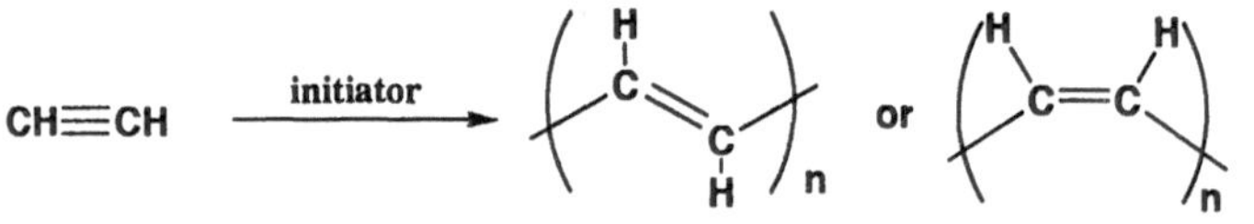

Scheme 25. Organolanthanide-initiated polymerization of acetylene

Table 13. Polymerization of acetylene derivatives

Monomer	Initiator	Polymer
$HC \equiv CH$	$Ln(naphthenate)_3/AliBu_3/Donor$ (Ln = Nd, Gd, La)	*cis*-content, 96%
$HC \equiv CH$	$Nd(OiPr)_3/AlEt_3$	
$RC \equiv CH$ ($R = C_5H_{11}, C_4H_9$)	$Nd(naphthenate)_3/AliBu_3$	*cis*-content, 90%
$C_6H_5C \equiv CH$	$La(naphthenate)_3/AliBu_3$	*cis*-content, 90% ($M_n = 2 \times 10^5$)
$C_6H_5C \equiv CH$	$Sc(naphthenate)_3/AliBu_3$	*cis*-content, 90% ($M_n = 0.8-3.2 \times 10^4$)

two exothermic peaks at 200 and 380 °C and an endothermic peak at 460 °C. These peaks were attributed to *cis-trans* isomerization, hydrogen migration, and chain decomposition, respectively [113].

Sc(naphthenate)$_3$/ROH/AlR$_3$ (1/2/7) has been found to exhibit an activity similar to the lanthanide series catalyst [114]. The *cis* PA film obtained with it showed an electrical conductivity of 14.4 S cm^{-1} when the polymer was doped with I$_2$ at a ratio of (CHI$_{0.04}$)$_n$, and the TEM measurement suggested the formation of ca. 20–30 nm fibrils.

The (P$_{204}$)$_3$Ln/AlR$_3$ system also exhibited good activity in the polymerization of acetylene when the Al/Ln ratio was 5 [114]. The polymerization was carried out conventionally, and polymers with a silver metallic appearance were obtained. The addition of an oxygen-containing donor effectively enhanced the polymerization rate and the *cis*-content. The effects were especially marked for P$_{204}$ (PO/Nd = 1.1). The activity decreased in the order Nd = Tb > Ce > Pr = Y > La > Er > Ho > S = Eu > Yb = Lu > Gd > Tm > Dy and the *cis*-content decreased in the order Pr (95%) > Lu = Tb = Dy (92%) > Er = Y = Sm = Gd (87–89%). The polymerization activities of Nd(P$_{507}$)$_3$, Nd(P$_{204}$)$_3$, and Nd(P$_{215}$) were compared and found to increase in this order; the result is consistent with the basicities of the ligands (P$_{507}$H = pKa 4.10, P$_{204}$H = pKa 3.32, P$_{215}$H = pKa 3.22). The M–C bond is assumed to weaken as the electron-donating ability of the ligand increases. The Nd(iPrO)$_3$/AlEt$_3$ (Al/Nd = 10) system [115] was also shown to be a good initiator for the polymerization of acetylene. The soluble fraction obtained was considered to be *trans*-polyacetylene, which was shown to have a molecular weight of 277–540. Its ^{1}H-NMR spectrum revealed methyl groups at δ = 0.826 ppm and terminal vinyl groups at 4.95 ppm.

Phenylacetylene was polymerized to produce a polymer of high *cis*-configuration using the Ln(naphthenate)$_3$/AlEt$_3$ system [116, 117]. The activity decreased in the order Gd > Lu > Nd = Ce > Ho > Sm > Dy = Eu > Er > Pr > La > Y = Tm > Yb, and the *cis*-content exceeded 90%. It had M_n and M_w of 2×10^5 and 4×10^5, respectively, and was crystalline according to XRD and SEM measurement. Its softening point was in the range 215–230 °C. Other terminal alkynes such as 1-hexyne, 1-pentyne, 3-methyl-1-pentyne, 4-methyl-1-pentyne, and 3-methyl-1-butyne, were found to polymerize quantitatively in *cis*-fashion with the Ln(naphthenate)$_3$/AlR$_3$/C$_2$H$_5$OH (Ln = Sc, Nd) or Ln(P$_{204}$)$_3$/AlR$_3$/C$_2$H$_5$OH (1:7:3) system. The highest molecular weight M_n obtained was 16.8×10^4 for poly(1-pentyne). Trimethylsilylacetylene was oligomerized to H(Me$_3$SiC = CH)$_n$CH$_2$CHMe$_2$ ($n = 2 - 3$) by using LnX$_3$(Donor)/AliBu$_3$ (Ln = Gd, Pr, Nd, Tb, Dy, Lu; X = Cl, Br) [118, 119]. The catalytic dimerization of terminal alkynes using (C$_5$Me$_5$)$_2$LnCH(SiMe$_3$)$_2$ (Ln = Y, La, Ce) has been reported recently. Here, the dimer was a mixture of 2,4-disubstituted 1-buten-3-yne and 1,4-disubstituted 1-buten-3-yne for phenylacetylene and (trimethylsilyl) acetylene, but it was a 2,4-disubstituted dimer for alkylacetylene [120]. Selective formation of 2,4-disubstituted 1-buten-3-ynes has already been achieved with the (C$_5$Me$_5$)$_2$TiCl$_2$/RMgX catalyst [121].

6 Block Copolymerization of Ethylene with Polar Monomers

Block copolymerization of ethylene or propylene with polar monomers has yet to be attained in polyolefin engineering. If succeeded, it should produce hydrophobic polymeric materials having remarkably high adhesive, dyeing, and moisture-adsorbing properties. The following is the first example of a well-controlled block copolymerization using the unique dual catalytic function of $LnR(C_5Me_5)_2$ (Ln = Sm, Yb, Lu; R = H, Me) complexes for both polar and nonpolar monomers [122]. Ethylene was copolymerized with MMA first by the homopolymerization of ethylene (17–20 mmol) with $SmMe(C_5Me_5)_2(THF)$ or $[SmH(C_5Me_5)_2]_2$(0.05 mmol) at 20 °C in toluene under atmospheric pressure and then by sequential addition of MMA (10 mmol) (Table 14). The initial step proceeded very rapidly (completion in 2 min) and produced a polymer with M_n = ca. 10 100 and M_w/M_n = 1.42–1.44. However, the second step was rather slow, with the reaction taking 2 h at 20 °C (Scheme 26). The polymer obtained was soluble in 1,2-dichlorobenzene and 1,2,4-trichlorobenzene at 100 °C but insoluble in THF and $CHCl_3$, indicating quantitative conversion to the desired linear block copolymer. Repeated fractionation in hot THF did not change the molar ratio of the polyethylene and poly(MMA) blocks, though poly(MMA) blended with polyethylene can be easily extracted with THF. With the copolymerization, the elution maximum in GPC shifted to a higher molecular weight region, with its initial unimodal pattern unchanged. The relative molar ratio of the polyethylene and poly(MMA) blocks was controllable at will in the range of 100:1 to 100:103 if the M_n of the initial polyethylene was fixed to ca. 10 300. ^{1}H- and ^{13}C-NMR spectra for the copolymers as well as their

Table 14. Block copolymerization of ethylene with polar monomers

Polar monomer	Polyethylene block[a]		Polar polymer block[b]		Unit ratio
	$M_n/10^3$	M_w/M_n	$M_n/10^3$	M_w/M_n	
MMA	10.3	1.42	24.2	1.37	100:103
	26.9	1.39	12.8	1.37	100:13
	40.5	1.40	18.2	1.90	100:12
MeA	6.6	1.40	15.0	1.36	100:71
	24.5	2.01	3.0	1.66	100:4
EtA	10.1	1.44	30.8	2.74	100:85
	24.8	1.97	18.2	3.84	100:21
VL	10.1	1.44	7.4	1.45	100:20
	24.8	1.97	4.7	1.97	100:5
CL	6.6	1.40	23.9	1.76	100:89
	24.5	2.01	6.9	2.01	100:7

Polymerization was carried out at 0 °C
[a] Determined by GPC using standard polystyrene
[b] Determined by ^{1}H-NMR

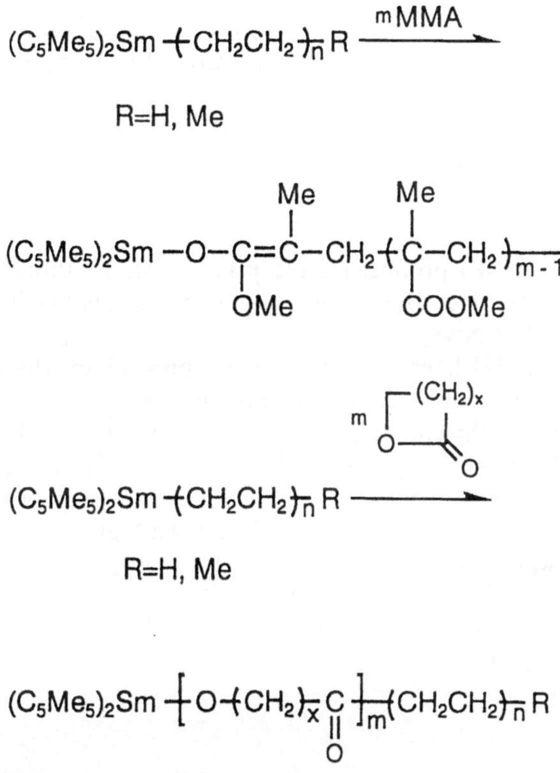

$$(C_5Me_5)_2Sm \; \text{+} CH_2CH_2 \text{+}_n R \quad \xrightarrow{\;mMMA\;}$$

R=H, Me

$$(C_5Me_5)_2Sm -O-\underset{\underset{OMe}{|}}{\overset{\overset{Me}{|}}{C}}=\underset{}{\overset{\overset{Me}{|}}{C}}-CH_2 \text{+} \underset{\underset{COOMe}{|}}{\overset{\overset{Me}{|}}{C}}-CH_2 \text{+}_{m-1} \text{+} CH_2CH_2 \text{+}_n R$$

$$(C_5Me_5)_2Sm \; \text{+} CH_2CH_2 \text{+}_n R \quad \xrightarrow{\quad m \overset{(CH_2)_x}{\underset{O}{\diagup\diagdown}}\quad}$$

R=H, Me

$$(C_5Me_5)_2Sm \; \text{+} O \text{+} CH_2 \text{+}_x \underset{\underset{O}{||}}{C} \text{+}_m (CH_2CH_2)_n R$$

Scheme 26. Block copolymerization of ethylene with MMA

IR absorption spectra were superimposable onto those of the physical mixtures of the respective homopolymers. The molar ratio of the poly(MMA) and polyethylene blocks, however, decreased as the M_n of the prepolymer increased, especially when it exceeded ca. 12 000 at which polyethylene began precipitating as fine colorless particles. It is noteworthy that smooth block copolymerization of ethyl acrylate or methyl acrylate to the growing polyethylene chain ($M_n = 6\,600$–$24\,800$) can be realized by the sequential addition of the two monomers.

Yasuda et al. [122] extended the above work to the block copolymerization of ethylene with lactones. δ-Valerolactone and ε-caprolactone were combined with the growing polyethylene end at ambient temperature and the expected AB-type copolymers (100 : 1 to 100 : 89) were obtained at high yield. Reversed addition of the monomers (first MMA or lactones and then ethylene) induced no block copolymerization at all, even in the presence of excess ethylene, and only homo-poly(MMA) and homo-poly(lactone) were produced.

The treatment of the resulting block copoly(ethylene/MMA) (100 : 3, $M_n = 35\,000$) and block copoly(ethylene/ε-caprolactone) (100 : 11, $M_n = 12\,000$)

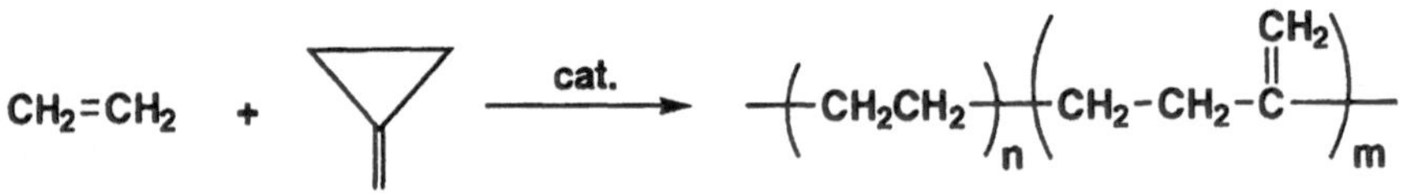

Scheme 27. Block copolymerization of ethylene with methylenecyclopropane

with dispersed dyes (Dianix AC-E) produced three primary colors, though polyethylene itself was inert to the dyes. Hence, these copolymers can be said to have a very desirable chemical reactivity.

More recently, Yang et al. [123] have examined a new approach in which a reactive functional group was introduced into polyolefins using methylenecyclopropane (Scheme 27). Thus, ethylene (1.0 atm) was copolymerized with methylenecyclopropane (0.25–2.5 ml) using $[LnH(C_5Me_5)_2]_2$ (Ln = Sm, Lu) in toluene at 25 °C. It was shown that 10–65 units of exo-methylenes were incorporated per 1000-CH_2- units and the resulting polymer had an M_w of $66–92 \times 10^3$, yet its M_w/M_n was > 4.

7 Conclusions

This article has reviewed recent developments in the rate earth metal-initiated polymerization of polar and nonpolar monomers. Most monomers, including alkyl methacrylates, alkyl acrylates, alkyl isocyanates, lactones, lactide, cyclic carbonates, ethylene, 1-olefin, conjugated diene, and acetylene derivatives, can be polymerized effectively with the help of the versatile function of rare earth metal initiators, except for such monomers as isobutene, vinyl ether, and vinyl pyridine. Their polymerization by a single-component initiator generally proceeds in living fashion and produces high molecular weight polymers $M_n > 400\,000$) with a very narrow molecular weight distribution ($M_w/M_n < 1.05$) at high conversion. Thus, the next problem to be solved is isotactic and syndiotactic polymerizations leading to polymers high enough in average molecular weight and narrow enough in molecular weight distribution. Stereospecific polymerization at high temperatures is especially important for alkyl (meth) acrylate, styrene, and 1-olefins from the industrial point of view. More sophisticated designs of ligands should be attempted. Although binary and ternary initiator systems are still in use for a variety of polymerizations, their replacement by single-component catalysts will be desirable in the near future.

Acknowledgement. The authors are grateful to Dr. Hiroshi Fujita, Emeritous Professor of Osaka University, for valuable discussion and comments.

8 References

1. Klein JW, Lamps JP, Gnaonou Y, Remp P (1991) Polymer 32: 2278.
2. (a) Webster OW, Hertler WR, Sogah DY, Farnham WB, RajanBabu TV (1983) J Am Chem Soc 105: 5706 (b) Sogah DY, Hertler WR, Webster OW, Cohen GM (1987) Macromolecules 20: 1473.
3. (a) Yasuda H, Yamamoto H, Yokota K, Miyake S, Nakamura A (1992) J. Am. Chem. Soc 114: 4908 (b) Yasuda H, Yamamoto H, Takemoto Y, Yamashita M, Yokota K, Miyake S, Nakamura A (1993) Makromol Chem Macromol Symp 67: 187 (c) Yasuda H, Yamamoto H, Yamashita, Yokota K, Nakamura A, Miyake S, Kai Y, Kanehisa N (1993) Macromolecules 22: 7134 (d) Yasuda H, Tamai H (1993) Prog Polym Sci 18: 1097 (e) Yasuda H, Ihara E (1996) The Polymeric Materials Encyclopedia, 10: 7359.
4. Ihara E, Morimoto M, Yasuda H (1995) Macromolecules 28: 7886.
5. (a) Jacobs C, Hershney SK, Hautekeer R, Fayt R, Jerome R, Teyssie Ph (1990) Macromolecules 23: 4025 (b) Hershney SK, Jacobs C, Hautekeer JP, Bayard Ph, Jerome R, Fayt R, Teyssie Ph (1991) Macromolecules 24: 4997 (c) Janata M, Müller AHE, Lochman L (1990) Makromol Chem 191: 2253 (d) Janata M, Lochman L, Vlecek P, Dybal J, Müller AHE (1992) Makromol Chem 101: 193.
6. Boffa LS, Novak BM (1994) Macromolecules 27: 6993.
7. Yu H, Choi W, Lim K, Choi S (1991) Macromolecules 24: 824.
8. Yamashita M, Takemoto Y, Ihara E, Yasuda H (1996) Macromolecules 29: 1798.
9. Urakawa O, Adachi K, Kotaka T, Takemoto Y, Yasuda H (1994) Macromolecules 27: 7410.
10. Stevels WM, Ankone MJK, Dijkstra PD, Feijin J (1996) Macromolecules 29: 3332.
11. Tanaka K, Ihara E, Yasuda H (1996) unpublished results.
12. Marks TJ, Ernst RD (1982) In: Wilkinson G, Stone FGA (ed) Comprehensive Organometallic Chemistry. Pergamon, Vol. 3, chap. 21.
13. Evans WJ, Drumond DK (1988) J Am Chem Soc 110: 2772.
14. Evans WJ, Drumond DK, Chamberlin LR, Doedens RJ, Bott SG, Zhang H, Atwood JL (1988) J Am Chem Soc 110: 4983.
15. Hatada K, Ute K, Tanaka K, Okamoto Y, Kitayama T (1986) Polym J 18: 1037.
16. Kitayama T, Shiozaki T, Sakamoto T, Yamamoto M, Hatada K (1989) Makromol Chem (suppl.) 15: 167.
17. (a) Szwarc M (1983) Adv Polym Sci 49: 1 (b) Nakahama S, Hirao A (1990) Prog Polym Sci 13: 299 (c) Inoue S (1988) Macromolecules 21: 1195.
18. (a) Sawamoto M, Okamoto O, Higashimura T (1987) Macromolecules 20: 2693. (b) Kojima K, Sawamoto M, Higashimura T (1988) Macromolecules 21: 1552.
19. (a) Gillon LR, Grubbs RH (1986) J Am Chem Soc 108: 733 (b) Grubbs RH, Tumas W (1989) Science 243: 907 (c) Schrock RR, Feldman J, Canizzo LF, Grubbs RH (1989) Macromolecules 20: 1169.
20. (a) Evans WJ, Grate JW, Choi W, Bloom I, Hunter WE, Atwood JL (1985) J Am Chem Soc 107: 941 (b) Evans WJ, Chamberlain LR, Ulibarri TA, Ziller JW (1988) J Am Chem Soc 110: 6423 (c) Evans WJ, Bloom I, Hunter WE, Atwood JL (1983) J Am Chem Soc 105: 1401.
21. Cao ZK, Okamoto Y, Hatada K (1986) Kobunshi Ronbunshu 43: 857.
22. (a) Joh Y, Kotake Y (1976) Macromolecules 3: 337 (b) Hatada K, Nakanishi H, Ute K, Kitayama T (1986) Polym J 18: 581.
23. Abe H, Imai K, Matsumoto M (1968) J Polym Sci C23: 469.
24. Nakano N, Ute K, Okamoto Y, Matsuura Y, Hatada K (1989) Polym J 21: 935.
25. Bawn CEH, Ledwith A (1962) Quarterly Rev 16: 361.
26. Cram DJ, Kopecky KR (1959) J Am Chem Soc 81: 2748.
27. Yamamoto Y, Giardello MA, Brard L, Marks TJ (1995) J Am Chem Soc 117: 3726.
28. Tokimitu T, Ihara E, Yasuda H, unpublished result.
29. Nitto Y, Hayakawa T, Ihara E, Yasuda H, unpublished result.
30. (a) Kuroki M, Watanabe T, Aida T, Inoue S (1991) J Am Chem Soc 113: 5903. (b) Aida T, Inoue S (1996) Acc Chem Res 29: 39.
31. Sun J, Wang G, Shen Z (1993) Yingyong Huaxue 10: 1.
32. Collins S, Ward DG, Suddaby KH (1994) Macromolecules 27: 7222.

33. Soga K, Deng H, Yano T, Shiono T (1994) Macromolecules 27: 7938.
34. Hosokawa Y, Kuroki M, Aida T, Inoue S (1991) Macromolecules 24: 8243.
35. (a) Ihara E, Morimoto M, Yasuda H (1996) Proc Japan Acad 71: 126. (b) Ihara E, Morimoto M, Yasuda H (1995) Macromolecules 28: 7886 (c) Suchoparek M, Spvacek J (1993) Macromolecules 26: 102 (d) Madruga EL, Roman JS, Rodriguez MJ (1983) J Polym Sci, Polym Chem Ed 21: 2739.
36. Ren J, Hu J, Shen Q (1995) Chinese J Appl Chem 12: 105.
37. Hu J, Qi G, Shen Q (1995) J Rare Earths 13: 144.
38. Kamide K, Ono H, Hisatani K (1992) Polymer J 24: 917.
39. (a) Patten TE, Novak BM (1991) J Am Chem Soc 113: 5065 (b) Patten TE, Novak BM (1993) Makromol Chem Macromol Symp 67: 203.
40. Fukuwatari N, Sugimoto H, Inoue S (1996) Macromol Rapid Commun 17: 1.
41. Wang J, Nomura R, Endo T (1996) J Polym Sci Polym Chem 33: 869.
42. Wang J, Nomura R, Endo T (1995) J Polym Sci Polym Chem 33: 2901.
43. (a) Hofman A, Szymanski R, Skomkowski S, Penczek S (1980) Makromol Chem 185: 655 (b) Agostini DE, Lado JB, Sheeton JR (1971) J Polym Sci A1: 2775.
44. Cherdran H, Ohse H, Korte F (1962) Makromol Chem 56: 187.
45. Yamashita M, Ihara E, Yasuda H (1996) Macromolecules 29: 1798.
46. Evans WJ, Katsumata H (1994) Macromolecules 27: 4011.
47. Shen Z, Chen X, Shen Y, Zheng Y (1994) J Polym Sci Polym Chem Ed 32: 597.
48. Akatsuka M, Aida T, Inoue S (1995) Macromolecules 28: 1320.
49. Shen Z, Sun J, Zhang Y (1994) Chinese Sci Bull 39: 1005.
50. Evans JE, Shreeve JL, Doedens RJ (1993) Inorg Chem 32: 245.
51. Shiomi M, Shirahama H, Yasuda H, unpublished result.
52. (a) Meerwein H, Kroning E (1987) J Pract Chem 147: 257 (b) Meerwein H (1947) Angew Chem 59: 168.
53. Li F, Jin Y, Pei F, Wang F (1993) J Appl Polym Sci 50: 2017.
54. Nomura N, Narita M, Endo T (1995) Macromolecules 28: 86.
55. (a) Nomura R, Endo T (1995) Polymer Bull 35: 683 (b) Nomura R, Endo T (1995) Macromolecules 28: 5372.
56. Zhang Y, Chen X, Shen Z (1989) Inorg Chimica Acta 155: 263.
57. Wu J, Shen Z (1990) J Polym Sci Polym Chem Ed 28: 1995.
58. Itoh H, Shirahama H, Yasuda H (1996) unpublished results.
59. Shen Z, Wu J, Wang G (1990) J Polym Sci Chem Ed 28: 1965.
60. Wu J, Shen Z (1990) Polymer J 22: 326.
61. (a) Inoue S, Tsuruta T (1969) J Polym Sci Polym Lett Ed 7: 287 (b) Kobayashi M, Inoue S, Tsurtuta T (1973) J Polym Sci Polym Chem Ed 11: 2383.
62. (a) Shen X, Zhang Y, Shen Z (1994) Chinese J Polym Sci 12: 28 (b) Shen Z, Chen X, Zhang Y (1994) Macromol Chem Phys 195: 2003.
63. (a) Machoon JP, Sigwalt P (1965) Compt Rend 260: 549 (b) Belonovskaja GP (1979) Europ Polym J 15: 185.
64. Dumas P, Spassky N, Sigwalt P (1972) Makromol Chem 156: 65.
65. Guerin PB, Sigwalt P (1974) Eur Polym J 10: 13.
66. Shen Z, Zhang Y, Pebg J, Ling L (1990) Sci China 33: 553.
67. Shen Z, Zhang Y (1994) Chinese Sci Bull 39: 717.
68. Shen Z, Sun J, Wu Li, Wu L (1990) Acta Chim Sin 48: 686.
69. Yao Y, Shen Q, private communication.
70. Shen Y, Zhang F, Zhang Y, Shen Z (1995) Acta Polymerica Sin 222.
71. Takemoto Y, Yasuda H, unpublished result.
72. Shen Y, Shen Z, Zhang Y, Yao K, private communication.
73. Jeske G, Shock LE, Swepstone PN, Schumann H, Marks TJ (1985) J Am Chem Soc 107: 8103.
74. (a) Yasuda H, Ihara E (1993) J Synth Org Chem Jpn 51: 931 (b) Yasuda H, Ihara E, Yoshioka S, Nodono M, Morimoto M, Yamashita M (1994) Catalyst Design for Tailor-made Polyolefins, Kodansha-Elsevier, p 237 (c) Nodono M, Ihara E, Yasuda H (1996) unpublished result.
75. Fu P, Mark TB (1995) J Am Chem Soc 117: 10747.
76. Mogstad AN, Waymouth RM (1992) Macromolecules 25: 2282.
77. Coughlin EB, Shapiro PJ, Bercaw JE (1992) Polym Prep 33: 1226.
78. Bryan E, Bercaw JE (1992) J Am Chem Soc 114: 7607.
79. Ihara E, Nodono M, Yasuda H, Kanehisa N, Kai Y (1996) Macromol Chem Phys 197: 1909.

80. Evans WJ, Ulibbari TA, Ziller JW (1988) J Am Chem Soc 110: 6877.
81. Evans WJ, Ulibbari TA, Ziller JW (1990) J Am Chem Soc 112: 219.
82. Yang M, Cha C, Shen Z (1990) Polymer J 22: 919.
83. Kobayashi E, Aida S, Aoshima S, Furukawa J (1994) J Polym Sci Polym Chem Ed 32: 1195.
84. Hayakawa T, Ihara E, Yasuda H (1995) 69th National Meeting of Chem Soc Jpn, 2B530.
85. Cheng XC, Shen Q (1993) Chinese Chem Lett 4: 743.
86. Hu J, Shen Q (1993) Cuihua Xuebao 11: 16.
87. Ishihara N, Seimiya T, Kuramoto M, Uoi M (1986) Macromolecules 19 2464.
88. Ishihara N, Kuramoto M, Uoi M (1988) Macromolecules 21: 3356.
89. Yu G, Chen W, Wang Y (1981) Kexue Tongabo 29: 412.
90. Ji X, Png S, Li Y, Ouyang J (1986) Scientia Sinica 29: 8.
91. Jin S, Guan J, Liang H, Shen Q (1993) J Catalysis (Cuihua Xuebao) 159: 14.
92. Hu J, Liang H, Shen Q (1993) J. Rare Earths 11: 304.
93. (a) Jin Y, Li F, Pei F, Wang F, Sun Y (1994) Macromolecules 27: 4397 (b) Li F, Jin Y, Pei K, Wang F (1994) J Macromol Sci Pure Appl Chem A31: 273.
94. (a) Jin Y, Li X, Sun Y, Ouyang J (1982) Kexue Tongbau 27: 1189 (b) Jin Y, Li X, Lin Y, Jin S, Shi E, Wang M (1989) Chin Sci Bull 34: 390.
95. Ricc G, Boffa G, Porri L (1986) Markomol Chem Lapid Commun 7: 355.
96. (a) Natta G, Porri L (1964) Polym Prepr 5: 1163 (b) Natta G, Porri (1966) Adv Chem Ser 52: 24.
97. Wang F, Sha R, Jin Y, Wang Y, Zheng Y (1980) Scientic Sinica 23: 172.
98. Li X, Sun Y, Jin Y (1986) Acta Chimica Sinica 44: 1163.
99. Hu Y, Ze L, Shen Q (1993) J Rare Earths 11: 304.
100. Lee DH, Wang JK, Ahn TO (1987) J Polym Sci Polym Chem Ed 25: 1407.
101. Lee DH, Ahn TO (1988) Polymer 29: 71.
102. Jenkins DK (1985) Polymer 26: 147.
103. Julemont M, Walckiers E, Wariam R, Teyssie P (1974) Makromol Chem 175: 1675.
104. Ji YL, Liu GD, Yu GQ (1989) J Macrimol Sci Chem A26: 405.
105. Yu G, Li Y, Qu Y, Liu X (1993) Macromolecules 26: 6702.
106. Zhu Y, Liu Y, Yu G, Li X (1994) China Synth Rubber Ind 17: 280.
107. (a) Yin Z, Yu G, Li Y, Qu Y (1993) Yinyong Huaxue 10: 5. (b) Li Y, Shufen P, Dawei X, Ouyang J (1985) Polym Commun 2: 111.
108. Bergbreiter DE, Chen LB, Chandran R (1985) Macromolecules 18: 1055.
109. Wang P, Jin Y, Pei F, Jing F, Sun Y (1994) Acta Polym Sin 4: 392.
110. Ito T, Shirakawa H, Ikeda S (1974) J Polym Sci Polym Chem Ed 12: 11.
111. Luttinger LB (1962) J Org Chem 27: 159.
112. Shen Z, Wang Z, Can Y (1985) Inorg Chimica Acta 110: 55.
113. Shen Z, Yang M, Shi M, Cai Y (1982) J Polym Sci Polym Chem Ed 20: 411.
114. Shen Z, Yu L, Yang M (1985) Inorg Chimica Acta 109: 55.
115. Hu X, Wang F, Zhao X, Yan D (1987) Chin J Polym sci 5: 221.
116. Zhao J, Yang M, Yuan Y, Shen Z (1988) Zhonggono Xitu Xuebao 6: 17.
117. Shen Z, Farona MF (1984) J Polym Sci Polym Chem Ed 22: 1009.
118. Shen Z, Farona MF (1983) Polym Bull 10: 298.
119. Mullagaliev IR, Mudarisova RK (1988) Izv Akad Nauk SSSR Ser Khim 7: 1687.
120. Heers HJ, Teuben JH (1991) Organometallics 10: 1980.
121. Akita M, Yasuda H, Nakamura A (1984) Bull Chem Soc Jpn 57: 480.
122. Yasuda H, Furo M, Yamamoto H, Nakamra A, Miyake S, Kibino N (1992) Macromolecules 25: 5115.
123. Yang Y, Seyam AM, Fuad PF, Marks TJ (1994) Macromolecules 27: 4625.

Editor: Professor Hiroshi Fujita
Received December 1996.

Polymerizations in Liquid and Supercritical Carbon Dioxide

Dorian A. Canelas[1] and Joseph M. DeSimone[2]
Department of Chemistry, CB #3290, Venable and Kenan Laboratories,
University of North Carolina at Chapel Hill, Chapel Hill, NC 27599, USA
[1] E-mail: dcanelas@email.unc.edu   [2] E-mail: Desimone@email.unc.edu

In the past few years, remarkable progress has been made in defining the scope and limitations of carbon dioxide (CO_2) as an inert polymerization medium. It has appear that CO_2 represents a viable solvent choice for a variety of propagation mechanisms including both chain growth and step growth polymerizations. When the environmental advantages of CO_2 are combined with its ability to be used as a solvent/dispersing medium for a wide variety of chemical reactions, it becomes clear that CO_2 may be the solvent of the future for the polymer industry. In addition, the design and synthesis of micelle-forming surfactants for CO_2 opens the doors for use of surfactant-modified CO_2 as the medium for heterogeneous polymerizations. This review will focus on the use of CO_2 as an inert solvent for the synthesis and processing of polymers.

List of Symbols and Abbreviations

AIBN	–	2,2′-azobis(isobutyronitrile)
ASB	–	anchor-soluble balance
BEMO	–	3,3′-bis(ethoxymethyl)oxetane
CFCs	–	chlorofluorocarbons
CO_2	–	carbon dioxide
D_4	–	octamethylcyclotetrasiloxane
FOA	–	1,1-dihydroperfluorooctyl acrylate
FOx-7	–	3-methyl-3′-[(1,1-dihydroheptafluorobutoxy)methyl]oxetane
5FSt	–	pentafluorostyrene
LCST	–	lower critical solution temperature
HDPE	–	high density polyethylene
HFP	–	hexafluoropropylene
HLB	–	hydrophilic-lipophilic balance
IBVE	–	isobutyl vinyl ether
MMA	–	methyl methacrylate
M_n	–	number average molecular weight
M_w	–	weight average molecular weight
MWD	–	molecular weight distribution
P_c	–	critical pressure
PAA	–	poly(acrylic acid)
PAN	–	poly(acrylonitrile)
PC	–	poly(carbonate)
PCTFE	–	poly(chlorotrifluoroethylene)
PDI	–	polydispersity index of the molecular weight distribution (M_w/M_n)
PDMS	–	poly(dimethylsiloxane)
PEMA	–	poly(ethyl methacrylate)
PEO	–	poly(ethylene oxide)
PET	–	poly(ethylene terephthalate)
PFOA	–	poly(1,1-dihydroperfluorooctyl acrylate)
PMA	–	poly(methyl acrylate)
PMMA	–	poly(methyl methacrylate)
PMP	–	poly(4-methyl-1-pentene)
PPO	–	poly(2,6-dimethylphenylene oxide)
PPVE	–	perfluoro(propyl vinyl ether)
PS	–	polystyrene
PVAc	–	poly(vinyl acetate)
PVC	–	poly(vinyl chloride)
PVF_2	–	poly(vinylidene fluoride)
PTFE	–	poly(tetrafluoroethylene)
STF	–	*p*-perfluoroethyleneoxymethylstyrene
T_c	–	critical temperature

T_g – glass transition temperature
TEMPO   – 2,2,6,6-tetramethyl-1-piperidinyloxy free radical
TFE   – tetrafluoroethylene
TMPCl   – 2-chloro-2,4,4-trimethylpentane
VF_2 – vinylidene fluoride
VOCs   – volatile organic compounds

1 Introduction

Recently, the scientific community has developed considerable interest in using liquid and supercritical CO_2 as a continuous phase for polymerization reactions. The polymer industry, in particular, is under increasing scrutiny to reduce emission of volatile organic compounds (VOCs), to completely phase out the use of chlorofluorocarbons (CFCs), and to reduce the generation of aqueous waste streams. Hence, these environmental concerns provide the principle driving force which is motivating the development of CO_2-based polymerization technologies. Unique properties such as tunable density and the ability to significantly plasticize glassy polymers make supercritical fluids interesting solvents which have been relatively unstudied. Since 1990, several groups have made a concerted effort to explore the scope and limitations of this fluid in polymer synthesis. This review will focus on the results of these recent efforts as well as the early work in this field which laid the groundwork for future investigations.

In general, the properties of supercritical fluids make them interesting media in which to conduct chemical reactions. A supercritical fluid can be defined as a substance or mixture at conditions which exceed the critical temperature (T_c) and critical pressure (P_c). One of the primary advantages of employing a supercritical fluid as the continuous phase lies in the ability to manipulate the solvent strength (dielectric constant) simply by varying the temperature and pressure of the system. Additionally, supercritical fluids have properties which are intermediate between those of a liquid and those of a gas. As an illustration, a supercritical fluid can have liquid-like density and simultaneously possess gas-like viscosity. For more information, the reader is referred to several books which have been published on supercritical fluid science and technology [1–4].

Savage and coworkers have recently published a review which summarizes many of the reactions which have been conducted at supercritical conditions in a variety of fluids [5]. In addition to reviewing some of the early work in polymerizations, their article gives a broad overview of the recent progress which has been made in both fundamental studies of the solvent properties of supercritical fluids as well as in applications ranging from electrochemistry to biomass utilization. In the arena of polymerizations, Scholsky reviewed polymerizations which occur at supercritical conditions, including those in which the system is above the critical conditions of the monomer, such as the high pressure process for the production of low density polyethylene [6]. Herein, on the other hand, we focus only on polymerizations which employ CO_2 as the inert continuous phase. This area is rapidly emerging as increasing evidence indicates that CO_2 constitutes the most logical solvent choice for a variety of polymerizations [7].

For numerous reasons, CO_2 represents an extremely attractive continuous phase in many applications. First of all, CO_2 is naturally occurring and readily available. Sources of CO_2 include both abundant natural reservoirs and recycled CO_2 which is recovered from the exhaust streams of power plants and industrial plants which produce ethanol, ammonia, hydrogen, and ethylene oxide [8]. Additionally, CO_2 has an easily accessible critical point with a T_c of 31.1 °C and a P_c of 73.8 bar [9]. Fig. 1 depicts the one component phase diagram for pure CO_2 [9]. Other benefits of the employment of CO_2 as a solvent include that it is inexpensive, non-flammable, non-toxic, and easily recycled. Moreover, it eliminates the need for energy intensive drying processes; the products from a polymerization conducted in CO_2 are isolated completely dry upon venting to remove the CO_2. Finally, liquid and supercritical CO_2 provide a relatively unexplored reaction medium which may give rise to new chemistry, especially in reactions which exhibit large solvent effects.

Because of its tunable density and low viscosity, synthetic organic chemists are beginning to utilize supercritical CO_2 as a medium for exploring reaction mechanisms and solvent cage effects [10, 11]. Asymmetric catalysis represents an area in which supercritical CO_2 may be useful as a solvent [12]. For polymerization reactions, in particular, the solvency of CO_2 as a medium and the plasticization effects of CO_2 on the resulting polymeric products represent the properties of central importance. These significant properties of CO_2 are explored in detail below. When all of these factors are combined with the fact that CO_2 may obviate the use of much more expensive and hazardous solvents,

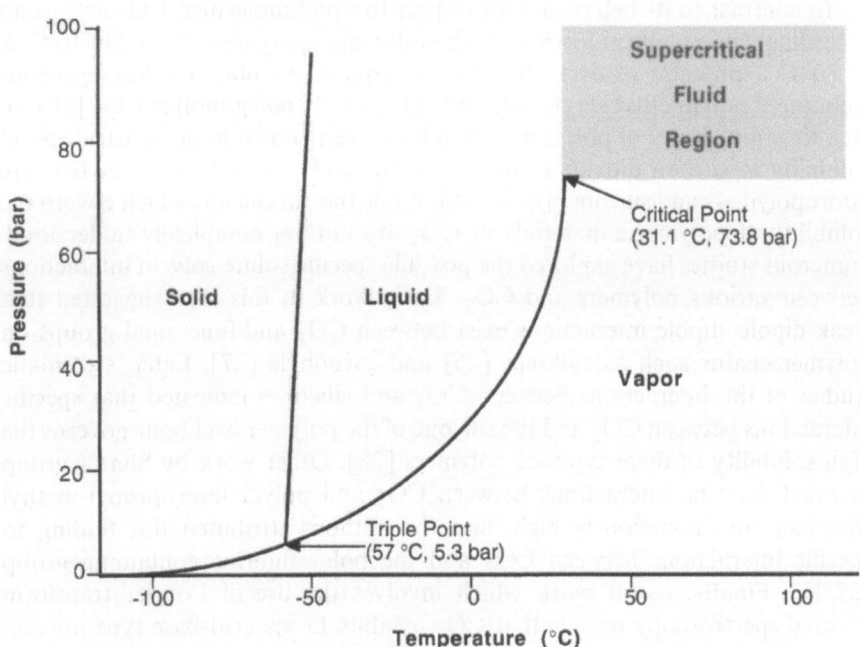

Fig. 1. One component phase diagram for pure CO_2

one cannot deny the importance of investigating its use as a continuous phase for polymer synthesis and processing.

When CO_2 is used as the continuous phase for a polymerization reaction, the solvency of the reagents and products is of primary importance. To begin with, CO_2 has a low dielectric constant; by varying temperature and density, Keyes and Kirkwood reported values ranging from 1.01 to 1.45 for gaseous CO_2 and 1.60–1.67 for liquid CO_2 [13]. The solubility parameter for CO_2, which is strongly dependent on pressure, has been calculated by McFann and coworkers [14]. It has been noted that CO_2 behaves very much like a hydrocarbon solvent with respect to its capability to dissolve small molecules, and thus many monomers exhibit high solubility in CO_2 [15]. Even though it is a relatively low dielectric media, CO_2 is a Lewis acid, and it has a strong quadrupole moment that allows it to dissolve some polar molecules such as methanol [16]. In contrast, other polar molecules such as amides, ureas, urethanes and azo dyes exhibit very poor solubility in CO_2 [15]. More importantly, the solubility of water was initially investigated by Lowry and Erickson who determined that less than 0.05 weight percent of water dissolves in liquid CO_2 over a range of temperatures [17]. King and coworkers have completed a more extensive study of the CO_2/water binary system at both liquid and supercritical conditions [18]. Bartle and coworkers have compiled a large table of the solubilities of compounds of low volatility in supercritical CO_2 [19]. Indeed, the solubility characteristics of compressed CO_2 with respect to small molecules has been extensively studied.

In contrast to its behavior with respect to small molecules, CO_2 acts as an exceedingly poor solvent for most high molar mass polymers. As an illustration, at 80 °C a pressure of over 2000 bar is required to obtain a homogeneous solution of poly(methyl acrylate) (PMA; $\langle M_n \rangle = 10\,600$ g/mol) in CO_2 [20]. In fact, the only classes of polymers which have been shown to demonstrate good solubility in carbon dioxide at mild conditions (T < 100 °C, P < 350 bar) are fluoropolymers and silicones [4, 21–25]. While the parameters which govern the solubility of polymeric materials in CO_2 are not yet completely understood, numerous studies have explored the possible specific solute-solvent interactions between various polymers and CO_2. Early work in this area suggested that weak dipole–dipole interactions exist between CO_2 and functional groups on polymer chains such as sulfones [26] and carbonyls [27]. Later, systematic studies of the interactions between CO_2 and silicones indicated that specific interactions between CO_2 and the silicone of the polymer backbone governs the high solubility of these types of polymers [28]. Other work by Stern's group revealed that the interactions between CO_2 and poly(trifluoropropyl methyl siloxane) are anomalously high, and the authors attributed this finding to specific interactions between CO_2 and the polar fluorine containing group [28,29]. Finally, recent work which involves the use of Fourier transform infrared spectroscopy reveals that CO_2 exhibits Lewis acid-base type interactions with electron donating functional groups of polymer chains such as the carbonyl group of poly(methyl methacrylate) (PMMA) [30]. Regardless of the nature of these interactions, the poor solubility of most polymers in CO_2 demands that the polymerization of most industrially important hydrocarbon monomers at reasonably accessible temperatures and pressures be conducted via heterogeneous polymerization techniques.

In addition to the solubility considerations, the plasticization of the resulting polymers by CO_2 constitutes another important issue which must be taken into account when polymerizations are conducted in liquid or supercritical CO_2. This plasticization of polymers plays a key role in both the diffusion of monomer into the polymer phase [31–33] as well as the incorporation of additives to a polymer matrix [34–37]. In 1982, Wang, Kramer, and Sachse studied the effects of high pressure CO_2 on the glass transition temperature and mechanical properties of polystyrene (PS) by measuring Young's modulus and creep compliance [38]. This work demonstrated the severe plasticization of PS by CO_2 as shown in Fig. 2 which depicts the T_g of PS as a function of CO_2 pressure [38]. In 1985, Chiou, Barlow, and Paul used differential scanning calorimetry to estimate the glass transition temperatures of polymers containing the dissolved gas [39]. In these experiments, the authors studied the plasticization of PMMA, PS, polycarbonate (PC), poly(vinyl chloride) (PVC), poly(ethylene terephthalate) (PET), and blends of PMMA and poly(vinylidene fluoride) (PVF_2) by CO_2 at pressures up to 25 bar, and reductions in T_g of up to 50 °C were observed. Wissinger and Paulaitis measured polymer swelling [40] and creep compliance as a function of time [41] to determine the T_gs of PC, PMMA, and PS in CO_2 at elevated pressures. Goel and Beckman have examined the plasticization of PMMA by CO_2 as a function of pressure using dielectric measurements [42]. Condo, Paul, and Johnston have completed an in-depth study of the effect of

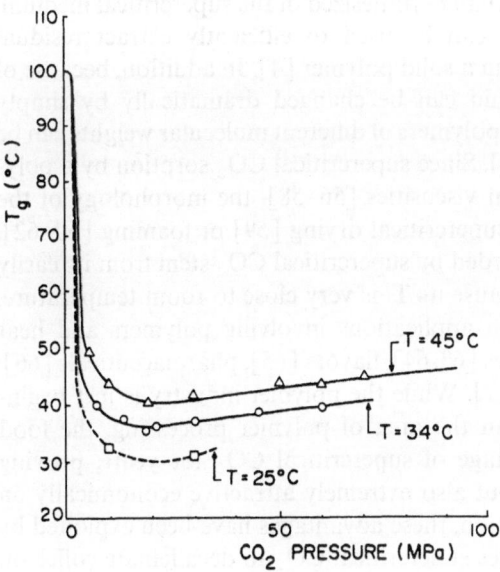

Fig. 2. T_g of polystyrene as a function of CO_2 pressure at different temperatures [38]

CO_2 solubility on T_g versus CO_2 partial pressure behavior for PMMA, PS, and a random copolymer of PMMA and PS [43]. Condo and Johnston have also used in situ measurements of creep compliance to determine the T_g depression of PMMA and poly(ethyl methacrylate) (PEMA) in CO_2 [44]. Kamiya and coworkers have studied the sorption and dilation of PEMA, poly(vinyl benzoate), and poly(vinyl butyral) [45, 46]. Handa and coworkers have studied the depression of the glass transition temperature of poly(2,6-dimethyl phenylene oxide) (PPO) by CO_2 at supercritical conditions [47]. Briscoe and Kelly have explored the plasticization of polyurethanes in CO_2 by monitoring the hydrogen bonding of the N–H group via infrared spectroscopy [48]. Finally, Shieh and coworkers have explored the interactions of supercritical CO_2 with nine different semi-crystalline polymers and eleven different amorphous polymers [49, 50]. By systematically examining both thermal and mechanical properties of polymers over a wide range of conditions, they concluded that both the morphology and polarity of the polymer play key roles in determining the absorption of CO_2 and subsequent plasticization. Recently, various models have been developed to predict the depression of the T_g of a polymer in the presence of compressed fluids. Johnston and coworkers developed their model using lattice fluid theory and the Gibbs–Di Marzio criterion [51]. In contrast, Kalospiros and Paulaitis took a molecular thermodynamic approach to develop their predictive model [52]. The predictions of T_g by these models are in reasonable agreement with the experimental data. Without a doubt, the plasticization of polymers by compressed CO_2 plays an important role in many of the polymerization processes which are being developed.

Research which began in the 1980s focused on the many advantages that supercritical CO_2 offers in polymer processing. These well developed processing methods for polymers in supercritical fluids can be taken advantage of most efficiently when the polymer is actually synthesized in the supercritical medium. For example, supercritical CO_2 can be used to efficiently extract residual monomer, solvent, or catalyst from a solid polymer [4]. In addition, because of the density of a supercritical fluid can be changed dramatically by simply altering the pressure, a mixture of polymers of different molecular weights can be efficiently fractionated [22, 53–55]. Since supercritical CO_2 sorption by a polymer reduces its melt and solution viscosities [56–58], the morphology of the polymer can be controlled with supercritical drying [59] or foaming [60–62]. Other processing advantages afforded by supercritical CO_2 stem from its easily accessible critical conditions. Because its T_c is very close to room temperature, supercritical CO_2 can be used in applications involving polymers and heat sensitive materials such as enzymes [63, 64], flavors [65], pharmaceuticals [66], and highly reactive monomers [67]. While the polymer industry is just beginning to harness this advantage in the area of polymer processing, the food industry has been taking advantage of supercritical CO_2 for years, proving that its use is not only feasible but also extremely attractive economically on a world wide scale. As an illustration, these advantages have been exploited by the coffee industry, which now uses supercritical CO_2 to decaffeinate coffee on an industrial scale; in the past, dichloromethane was the solvent employed in this process [68, 69]. In fact, more than 90% of supercritical fluid separations performed today use CO_2 as the eluent [70]. Without a doubt, the many polymer processing advantages provided by CO_2 have just begun to be explored.

2 Homogeneous Solution Polymerizations

2.1 Fluoropolymers

Poor solubility in most common organic solvents represents an inherent problem in the synthesis and processing of many high molar mass fluoropolymers. In fact, CFCs and carbon dioxide are the best solvents for amorphous varieties of fluoropolymers. Due to the environmental problems associated with CFCs, the international community is seeking to replace them with more benign compounds such as hydrochlorofluorocarbons and hydrofluorocarbons. However, the environmental problems which will be created by the use of these replacement compounds such as the accumulation of trifluoroacetic acid in the atmosphere clouds this issue [71]. Carbon dioxide presents an ideal inert solvent to effect the polymerization of these types of highly fluorinated monomers and obviates the use of solvents that are being phased out because of environmental concerns.

2.1.1 Free Radical Chain Growth

Using supercritical CO_2 as the solvent, we have used free radical initiators to effect the synthesis of high molar mass amorphous fluoropolymers [23, 72–74]. Due to the high solubility of these polymers in the CO_2 continuous phase, the polymerizations remained homogeneous throughout the course of the reactions. Several fluorinated acrylate monomers have been polymerized using this methodology to give high yields of polymer. For example, the polymerization of 1,1-dihydroperfluorooctyl acrylate (FOA) is shown in Scheme 1. Other fluorinated acrylate polymers which have been synthesized in this manner include poly[2-(N-ethylperfluorooctanesulfonamido)ethyl acrylate], poly[2-(N-ethyl-perfluoro-octanesulfonamido)ethyl methacrylate], and poly[2-(N-methylper-fluorooctane-sulfonamido)ethyl acrylate] [75]. Styrenes with a perfluoroalkyl side chains in the para position, such as p-perfluoroethyleneoxymethylstyrene (STF), constitute a second type of monomer polymerized via homogeneous solution polymerization in supercritical CO_2 [72]. The synthesis and subsequent polymerization of STF is shown in Scheme 2. The product obtained from this polymerization in CO_2 was identical to the product of a solution

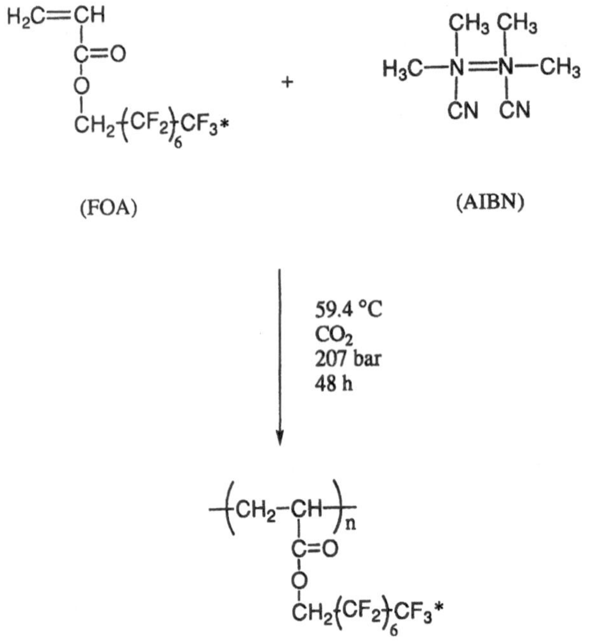

H₂C=CH
  |
  C=O
  |
  O
  |
  CH₂-(CF₂)₆-CF₃*

                    +

          CH₃ CH₃
           | |
H₃C—N=N—CH₃
           | |
          CN   CN

        (FOA)   (AIBN)

59.4 °C
CO₂
207 bar
48 h

-(CH₂-CH)ₙ-
       |
       C=O
       |
       O
       |
       CH₂-(CF₂)₆-CF₃*

Scheme 1. Homopolymerization of FOA in CO_2. * Contains ca. 25% $-CF_3$ branches [23]

polymerization in 1,1,2-trichloro-1,2,2-trifluoroethane (Freon-113), indicating that CO_2 does indeed represent an exceptionally good replacement solvent for this type of reaction in a CFC. Moreover, this technique has proven to be valuable in the synthesis of statistical copolymers of the appropriate fluorinated monomers and other hydrocarbon monomers such as methyl methacrylate (MMA), butyl acrylate, ethylene, and styrene (see Table 1). Homogeneous solution polymerization has also been employed to prepare CO_2-soluble polymeric amines by copolymerizing FOA with 2-(dimethylamino)ethylacrylate or 4-vinylpyridine [76].

We have also investigated the kinetics of free radical initiation using azobisisobutyronitrile (AIBN) as the initiator [24]. Using high pressure ultraviolet spectroscopy, it was shown that AIBN decomposes slower in CO_2 than in a traditional hydrocarbon liquid solvent such as benzene, but with much greater efficiency due to the decreased solvent cage effect in the low viscosity supercritical medium. The conclusion of this work was that CO_2 is inert to free radicals and therefore represents an excellent solvent for conducting free radical polymerizations.

The polymerization and oligomerization of fluorolefins in CO_2 has particular advantages over other solvents due to the lack of chain transfer to CO_2. The radical derived from fluorolefin monomers such as TFE are highly electrophilic and will chain transfer to almost any hydrocarbon present. In addition, highly reactive monomers such as TFE may be handled more safely as a mixture with

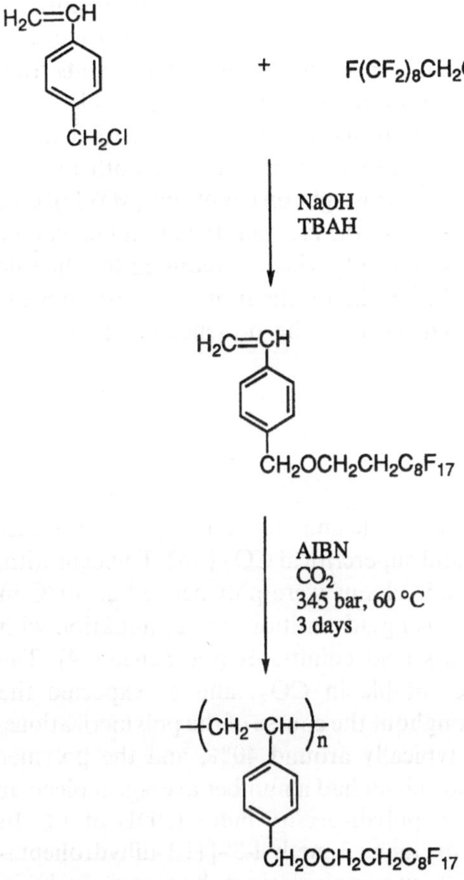

$H_2C=CH$

+ $F(CF_2)_8CH_2CH_2-OH$

NaOH
TBAH

$H_2C=CH$

$CH_2OCH_2CH_2C_8F_{17}$

AIBN
CO_2
345 bar, 60 °C
3 days

$+CH_2-CH+_n$

$CH_2OCH_2CH_2C_8F_{17}$

Scheme 2. Synthesis and polymerization of STF [72]

Table 1. Statistical copolymers of FOA with vinyl monomers. Polymerizations were conducted at 59.4 ± 0.1 °C and 345 ± 0.5 bar for 48 hours in CO_2. Intrinsic viscosities were determined in 1,1,2-trifluorotrichloroethane (Freon-113) at 30 °C [23]

Copolymer	Feed Ratio	Incorporated	Intrinsic Viscosity (dL/g)
poly(FOA-co-MMA)	0.47	0.57	0.10
poly(FOA-co-styrene)	0.48	0.58	0.15
poly(FOA-co-BA)	0.53	0.57	0.45
poly(FOA-co-ethylene)	0.35	–	0.14

CO_2 [67]. Early work examining the free-radical telomerization of TFE in supercritical CO_2 took advantage of the high solubility of the low molecular weight perfluoroalkyl iodides products to prepare these materials homogeneously [77]. In these reactions, we utilized perfluorobutyl iodide as the telogen and reactions were run both in the presence and absence of the thermal initiator AIBN. In the presence of AIBN, the results were found to be inconsistent and irreproducible, presumably due to chain transfer to initiator. The results from the experiments conducted in the absence of initiator are shown in Table 2, and a general telomerization of TFE in CO_2 is shown in Scheme 3. As Table 2 illustrates, the perfluoroalkyl iodides prepared in this manner had both control-led molecular weights and narrow molecular weight distributions (MWD). One of the major advantages of this process lies in the fact that the monomer can be used directly without prior removal of the CO_2 which is required for the safe shipping and handling of TFE. In addition, the results from these experiments show that CO_2 acts as an inert solvent even in the presence of highly elec-trophilic fluorocarbon radicals.

2.1.2 Cationic Chain Growth

We have polymerized several fluorinated cyclic and vinyl ethers via homogene-ous cationic polymerization in liquid and supercritical CO_2 [78]. To begin with, two vinyl ethers bearing fluorinated side chains were polymerized at 40 °C in a supercritical CO_2 continuous phase using adventitious water initiation with ethylaluminum dichloride as the Lewis acid coinitiator (see Scheme 4). The monomers and the coinitiator were soluble in CO_2, and as expected the solutions remained homogeneous throughout the course of the polymerizations. Conversion of these reactions were typically around 40%, and the polymer bearing the fluorinated sulfonamide side chain had a number average molecular weight (M_n) of 4.5×10^3 g/mol with a polydispersity index (PDI) of 1.6. In another experiment, a fluorinated oxetane, 3-methyl-3'-[(1,1-dihydrohepta-fluorobutoxy)methyl]oxetane (FOx-7), was polymerized homogeneously in liquid CO_2 at 0 °C and 289 bar for 4 h (See Scheme 5). In this reaction,

Table 2. Experimental results from the telomerizations of TFE in supercritical CO_2. Perfluorobutyl iodide was employed as the telogen [77]

[monomer]/[telogen]	yield (%)	M_n (g/mol)	PDI
1.6	88	570	1.35
1.5	87	590	1.38
1.8	86	630	1.38
2.2	78	650	1.44

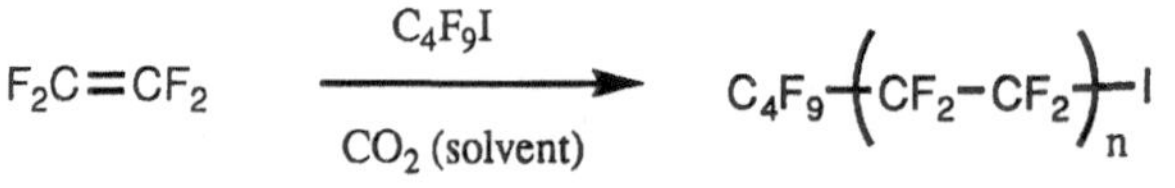

$$F_2C=CF_2 \quad \xrightarrow[\text{CO}_2 \text{ (solvent)}]{\text{C}_4\text{F}_9\text{I}} \quad C_4F_9 \left(CF_2-CF_2 \right)_n I$$

Scheme 3. Telomerization of TFE in supercritical CO_2 [77]

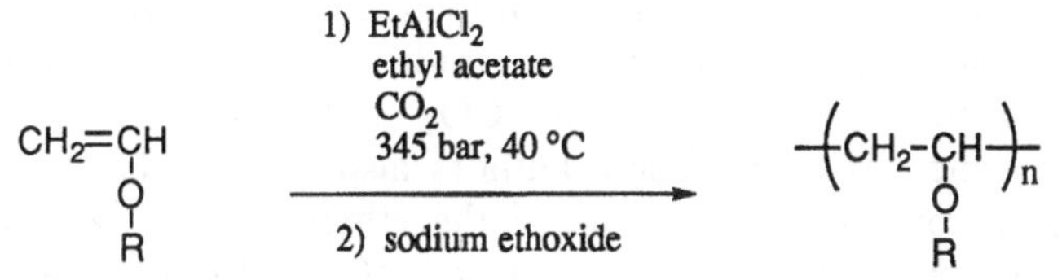

$$CH_2=CH \quad \xrightarrow[\text{2) sodium ethoxide}]{\substack{\text{1) EtAlCl}_2 \\ \text{ethyl acetate} \\ CO_2 \\ \text{345 bar, 40 °C}}} \quad -(CH_2-CH)_n-$$

(vinyl ether monomer with O–R substituent on left; polymer with O–R substituent on right)

R = -CH$_2$CH$_2$(CF$_2$)$_n$CF$_3$; n = 5-7

R = -CH$_2$CH$_2$NSO$_2$C$_8$F$_{17}$
                |
               C$_3$H$_7$

Scheme 4. Homogeneous polymerization of fluorinated vinyl ethers in supercritical CO$_2$ [78]

$$\text{(oxetane, FOx-7)} \quad \xrightarrow[\text{2) NaOH}]{\substack{\text{1) BF}_3\text{-THF} \\ CO_2 \\ \text{289 bar, 0 °C}}} \quad H-\left(O-CH_2-\underset{CH_2OCH_2CF_2CF_2CF_3}{\overset{\overset{\displaystyle CH_3}{|}}{\underset{|}{C}}}-CH_2\right)_n-OH$$

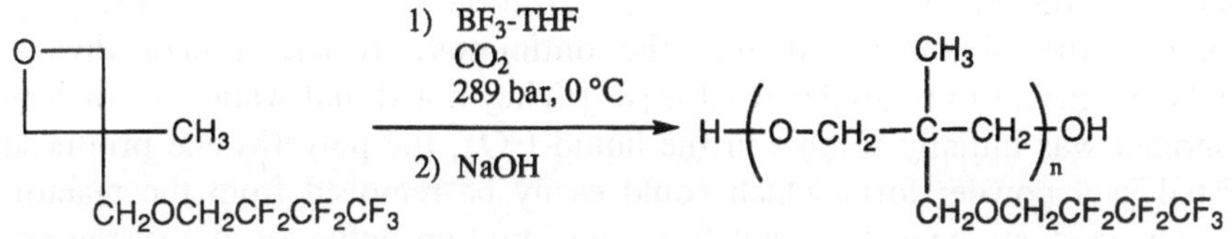

(FOx-7)

Scheme 5. Homogeneous cationic polymerization of FOx-7 in supercritical CO$_2$ [78]

trifluoroethanol was used as the initiator and boron trifluoride tetrahydrofuran-
ate as the coinitiator. The polymerizations resulted in a 77% yield of polymer
with a M$_n$ of 2.0×10^4 g/mol and a PDI of 2.0. This result for the oxetane
polymerization in CO$_2$ was comparable to a control experiment which was
conducted in Freon-113, another indication that CO$_2$ provides an excellent
alternative to CFCs in solvent applications. These results also indicate that
cationic chain growth can effectively be used in the homogeneous polymerization of fluorinated monomers in CO$_2$.

3 Heterogeneous Polymerizations

3.1 Free Radical Precipitation Polymerizations

Precipitation polymerizations dominated the early work which aimed at preparing industrially important hydrocarbon polymers in CO$_2$. In 1968, Hagiwara
and coworkers explored the polymerization of ethylene in CO$_2$ using both
gamma radiation and AIBN as free radical initiators [79]. Reactions were
conducted at pressures of 440 bar and over the temperature range of 20–45 °C.

CO_2 was chosen to be studied as a solvent for these polymerizations because of its high stability to ionizing radiation. In these studies, the infrared spectra of the polymers revealed that the presence of the CO_2 continuous phase had little effect on the polymer structure. These authors also studied the kinetics of initiation, propagation, and termination of the ethylene polymerizations in liquid CO_2 which were gamma radiation induced [80, 81]. In these studies, the authors determined that the rate of initiation increased with increasing CO_2 concentration because of the effects of the electron density of the reaction mixture on the absorption of the gamma radiation energy. In addition, while the CO_2 was shown to be inert to the growing oligomeric radicals at low dose rates $(9.0 \times 10^2$ rad/h), the oxygen which is produced at high dose rates $(2.5 \times 10^4$ rad/h) causes termination and the formation of carbonyl groups in the polymer. To extend this work and further investigate electron density and the role of reactive additives, the authors also studied the effects of various alkyl halides on the radiation induced polymerization of ethylene in CO_2 [82]. Moreover, they obtained a patent for the continuous polymerization of ethylene in CO_2 by means of a tubular reactor [83]. They noted that while the ethylene monomer was initially soluble in the liquid CO_2, the polyethylene produced existed in a powder form which could easily be removed from the reactor. Powder products typically result from precipitation polymerizations; the advantage of using CO_2 is the dryness of the resulting polymer.

In 1968, a French Patent issued to the Sumitomo Chemical Company disclosed the polymerization of several vinyl monomers in CO_2 [84]. The United States version of this patent was issued in 1970, when Fukui and coworkers demonstrated the precipitation polymerization of several hydrocarbon monomers in liquid and supercritical CO_2 [85]. As examples of this methodology, they demonstrated the preparation of the homopolymers PVC, PS, poly(acrylonitrile) (PAN), poly(acrylic acid) (PAA), and poly(vinyl acetate) (PVAc). In addition, they prepared the random copolymers PS-co-PMMA and PVC-co-PVAc. In 1986, the BASF Corporation was issued a Canadian Patent for the preparation of polymer powders through the precipitation polymerization of monomers in carbon dioxide at superatmospheric pressures [86]. Monomers which were polymerized as examples in this patent included 2-hydroxyethylacrylate and N-vinylcarboxamides such as N-vinyl formamide and N-vinyl pyrrolidone.

In 1988, Terry and coworkers attempted to homopolymerize ethylene, 1-octene, and 1-decene in supercritical CO_2 [87]. The purpose of their work was to increase the viscosity of supercritical CO_2 for enhanced oil recovery applications. They utilized the free radical initiators benzoyl peroxide and tert-butyl-peroctoate and conducted polymerization for 24–48 h at 100–130 bar and 71 °C. In these experiments, the resulting polymers were not well studied, but solubility studies on the products confirmed that they were relatively insoluble in the continuous phase and thus were not effective as viscosity enhancing agents. In addition, α-olefins are known not to yield high polymer using free radical methods due to extensive chain transfer to monomer.

The precipitation polymerization of acrylic acid has been well studied. An advantage of this process lies in the extremely fast propagation rate of this reaction which allows the synthesis of high molecular weight PAA even though it demonstrates poor solubility in the continuous phase. In 1986, BASF Corporation was issued a Canadian patent for this process which cited several different reaction conditions which all resulted in the successful preparation of PAA in a carbon dioxide continuous phase [88]. In 1989, a European patent application filed by B.F. Goodrich was published which described the preparation of acrylic acid type thickening agents in CO_2 [89]. We have explored the synthesis of PAA in supercritical carbon dioxide and have expanded this study to include effective molecular weight control through the use of ethyl mercaptan as a chain transfer agent [90]. Other work in this area by Dada and coworkers has shown that the molecular weight of the PAA product can be controlled by manipulation of the temperature and pressure [91].

We have recently reported the copolymerizations of tetrafluoroethylene (TFE) with perfluoro(propyl vinyl ether) (PPVE) and with hexafluoropropylene (HFP) in supercritical CO_2 [92,93]. In these precipitation polymerizations, bis(perfluoro-2-propoxy propionyl)peroxide was used as the free radical initiator and reactions were conducted at 35 °C and pressures below 133 bar. Good yields of high molar mass ($>10^6$ g/mol) copolymers resulted from these polymerizations. This CO_2-based system offers three primary advantages over the conventional aqueous process. First of all, the highly electrophilic monomers derived from fluorolefin monomers do not chain transfer to the CO_2 continuous phase. The second advantage lies in the fact that the aqueous process results in the production of materials with unstable and undesirable carboxylic acid and acid fluoride end-groups which can cause processing and performance problems. In the CO_2-based polymerization process, these deleterious end groups do not form to a significant extent. The third advantage lies in the safety issues involved in the shipping and handling of TFE as a mixture with CO_2 [67]; conducting the polymerizations directly in a CO_2 continuous phase eliminates the need for its removal.

Finally, the use of stable free radical polymerization techniques in supercritical CO_2 represents an exciting new topic of research. Work in this area by Odell and Hamer involves the use of reversibly terminating stable free radicals generated by systems such as benzoyl peroxide or AIBN and 2,2,6,6-tetramethyl-1-piperidinyloxy free radical (TEMPO) [94]. In these experiments, styrene was polymerized at a temperature of 125 °C and a pressure of 240–275 bar CO_2. When the concentration of monomer was low (10% by volume) the low conversion of PS which was produced had a M_n of about 3000 g/mol and a narrow MWD (PDI < 1.3). NMR analysis showed that the precipitated PS chains are primarily TEMPO capped, and the polymer could be isolated and then subsequently extended by the addition of more styrene under an inert argon blanket. The authors also demonstrated that the chains could be extended by keeping the precipitated PS in the CO_2 continuous phase and increasing the monomer concentration in the reactor to $\sim 30\%$ by volume.

3.2 Dispersion and Emulsion Polymerizations

Because the terminology to describe heterogeneous polymerization processes has been used inconsistently in the literature, a brief treatment of this subject is necessary to avoid confusion. In a heterogeneous process, either the monomer or the resulting polymer is insoluble in the continuous phase. The four common heterogeneous processes of suspension, emulsion, dispersion, and precipitation polymerization can be clearly distinguished on the basis the initial state of the polymerization mixture, the kinetics of polymerization, the mechanism of particle formation, and the shape and size of the final polymer particles [95]. Precipitation polymerizations which have been conducted in CO_2 are reviewed in Sect. 3.1. Dispersion and emulsion polymerization constitute the other two heterogeneous polymerization methods which have been explored using CO_2, and a brief discussion of the traditional definitions of these colloid forming processes follows.

In an emulsion polymerization, the reaction mixture is initially heterogeneous due to the poor solubility of the monomer in the continuous phase. In order for a reaction to take advantage of the desirable Smith–Ewart kinetics [96], the monomer and initiator must be segregated with the initiator preferentially dissolved in the continuous phase and not the monomer phase. Because of the kinetics of an emulsion polymerization, high molecular weight polymer can be produced at high rates. The polymer which results from an emulsion polymerization exists as spherical particles typically smaller than one μm in diameter. However, due to the high solubility of most vinyl monomers in CO_2, emulsion polymerization in CO_2 probably will not be a very useful process for commercially important monomers.

In contrast, a dispersion polymerization begins homogeneously due to the solubility of both the monomer and the initiator in the continuous phase. Once the growing oligomeric radicals reach a critical molecular weight, phase separation occurs. At this point the polymer is stabilized as a colloid, and as a result the polymerization continues to higher degrees of polymerization than the analogous precipitation reaction in the absence of stabilizer. Since the initiator and monomer are not segregated or compartmentalized, dispersion polymerizations do not follow Smith–Ewart kinetics; however, enhanced rates of polmerization are often observed due to the Gel effect within a growing polymer particle. The product from a dispersion polymerization also exists as spherical polymer particles, but these typically range in size from 100 nm to 10 μm [95]. Due to the good solubility of many small organic molecules in CO_2, dispersion polymerization constitutes the best method which has been developed thus far for producing high molecular weight, insoluble, industrially important hydrocarbon polymers.

3.2.1 Design and Synthesis of Stabilizers

The conventional methods used to prevent coagulation or flocculation of the polymer particles in a colloidal dispersion include electrostatic, electrosteric, and steric stabilization. Comprehensive reviews of these mechanisms can be found elsewhere in the literature [97–99]. Because of the latitude it allows in reaction conditions, steric stabilization offers several advantages over the other two mechanisms; a primary benefit lies in the effectiveness of polymeric stabilizers in solvents with low dielectric constants. For this reason, steric stabilization provides the stabilization mechanism of choice for CO_2 systems. When steric stabilization acts effectively in a heterogeneous system, the stabilizing molecule attaches to the surface of the polymer particle by either grafting or physical adsorption. Thus, the polymeric stabilizer is a macromolecule which preferentially exists at the polymer-solvent interface and prevents aggregation of particles by coating the surface of each particle and imparting long range repulsions between them. These long range repulsions must be great enough to compensate for the long range van der Waals attractions [98]. Peck and Johnston have developed a lattice fluid self-consistent field theory to describe a surfactant chain at an interface in a compressible fluid, allowing traditional colloidal stabilization theory [97] to be extended to supercritical fluid continuous phases [100, 101].

In traditional aqueous systems, it has long been recognized that the best polymeric stabilizers are those which are amphiphilic in nature. The term hydrophilic-lipophilic balance (HLB) has been used extensively in the literature to describe empirically the relative solubilities of the contrasting portions of the molecules in aqueous and organic media, respectively. For CO_2 systems, however, use of the term HLB would be inappropriate because of the unique solvation properties which it possesses. A more suitable term to describe the amphiphilic character of stabilizers for use in CO_2 is anchor-soluble balance (ASB). Qualitatively, ASB expresses the relative proportions of the soluble and insoluble components of the stabilizer [98]. A delicate balance exists between the soluble and insoluble segments; for the stabilizer to function effectively, it must possess an appropriate ASB. An amphiphilic polymer contains a segment which has high solubility in carbon dioxide, the "CO_2-philic" [102, 103] segment, as well as an anchoring segment which is relatively insoluble in CO_2, the "CO_2-phobic" segment [104]. The CO_2-phobic segment may be either hydrophilic or lipophilic, depending on the nature of the monomer being polymerized. A variety of compositions and architectures can be exploited when designing polymeric stabilizers for use in CO_2, including homopolymers as well as statistical, block, and graft copolymers.

Most commercially available surfactants were designed for use in an aqueous continuous phase, and thus they are completely insoluble in CO_2 as demonstrated by Consani and Smith who have studied the solubility of over 130 surfactants in CO_2 at 50 °C and 100–500 bar [105]. They concluded that microemulsions of commercial surfactants form much more readily in other low polarity supercritical fluids such as alkanes and xenon rather than in CO_2. They also noted the apparent compatibility of fluorinated hydrocarbons and CO_2, a fact which had been suggested by previous studies on perfluorinated polyethers by McHugh and Krukonis [4].

Several groups have designed and synthesized polymeric materials for the purpose of stabilizing the polymerization of water soluble monomers or dispersing large amounts of water into CO_2. Two approaches to this problem have been explored: (1) the synthesis of block and graft copolymers containing a hydrophilic polymer as one of the segments and (2) the synthesis of polymeric materials containing an ionic endgroup.

The first approach involved the design and synthesis of CO_2-philic/hydrophilic amphiphilic block and graft copolymers for use in the dispersion of hydrophilic molecules in CO_2. In 1993, we reported the use of the macromonomer technique to synthesize and characterize an amphiphilic graft copolymer with a CO_2-philic poly(1,1-dihydroperfluorooctyl acrylate) (PFOA) backbone and hydrophilic poly(ethylene oxide) (PEO) grafts [106]. In this study, solvatochromic characterization was employed to demonstrate that the PEO grafts enabled the solubility of the hydrophilic dye methyl orange in supercritical CO_2. This graft copolymer was further characterized by small angle X-ray scattering and shown to form spherical micelles in the presence of water in a CO_2 continuous phase [107]. This discovery constituted an important step in the right direction, since amphiphilic molecules which are effective stabilizers in heterogeneous polymerizations often form micelles in solution. Moreover, this work represented the first direct confirmation that micelles can form in a CO_2 continuous phase. Indeed, this strategy for the use of surfactant modified CO_2 could allow many current industrial processes, ranging from polymerizations to precision cleaning, to be converted from organic and aqueous based solvent systems to more environmentally benign CO_2.

In the second approach, investigators have taken advantage of the high solubility of fluorinated and siloxane based polymeric materials to transfer an ionic headgroup into CO_2. Due to the very low dielectric constant and polarizability of CO_2, it was anticipated that the ionic headgroups would associate in the presence of polar molecules such as water and charged metals and thus allow for polar materials to be uniformly dispersed in the non-polar continuous phase. For this purpose, Beckman and coworkers have studied the phase behavior of both silicone-based and fluoroether-functional amphiphiles in supercritical CO_2 [25, 108]. In these studies, the effect of the polarity of the hydrophilic headgroup on the resulting phase behavior of the surfactants was explored. In addition, the fluoroether-functional amphiphile was shown to permit the extraction of thymol blue from aqueous solution into CO_2. Another advance in this area was made when Johnston and coworkers demonstrated the formation of a one-phase microemulsion consisting of the hybrid fluorocarbon/hydrocarbon surfactant $C_7F_{15}CH(OSO_3^-Na^+)C_7H_{15}$ and water in CO_2 [109]. In this work, the water-to-surfactant ratio in a single phase microemulsion was as high as 32 at 25 °C and 231 bar. It was shown that the use of this type of surfactant increased the amount of water dissolved in the CO_2 by an order of magnitude. In more recent work, Johnston and coworkers have demonstrated the formation of aqueous microemulsion droplets in a CO_2 continuous phase using an ammonium carboxylate perfluoropolyether surfactant [110]. Several spectroscopic techniques were employed to investigate the properties of these aqueous microemulsions. This type of approach to the formation of microemulsions in non-polar supercritical fluids has been the focus of two recent reviews [111, 112].

Macromolecules have also been specifically designed and synthesized for use as emulsifiers for lipophilic materials and as stabilizers in the colloidal dispersion of lipophilic, hydrocarbon polymers in CO_2. We have demonstrated the amphiphilicity of fluorinated acrylate homopolymers, such as PFOA, which contain a lipophilic, acrylate like backbone and CO_2-philic, fluorinated side chains (see Fig. 3) [103]. It has been demonstrated that a homopolymer which physically adsorbs to the surface of a polymer colloid prevents agglomeration by the presence of loops and tails (see Fig. 4) [113]. The synthesis of this type of

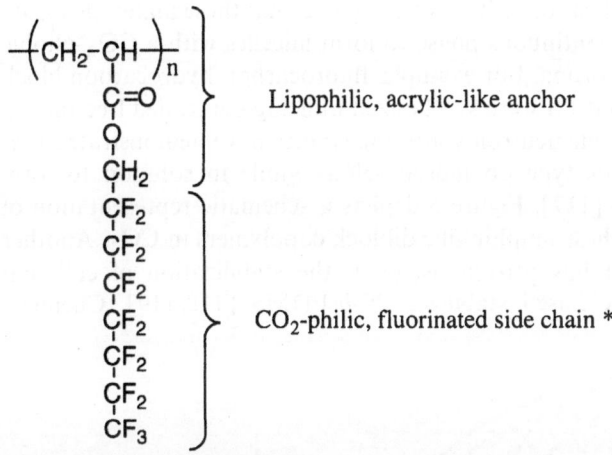

Fig. 3. Structure of amphiphilic polymeric stabilizer, PFOA. * Contains ca. 25% $-CF_3$ branches [103]

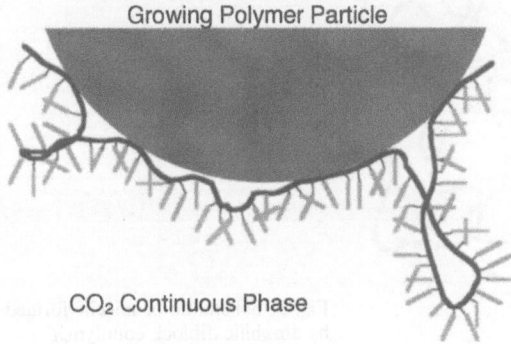

Fig. 4. Schematic of polymer colloidal particles stabilized by PFOA homopolymer [103]

fluorinated acrylate polymer via homogeneous solution polymerization in CO_2 was discussed in Sect. 2.1.1. The phase behavior [33] and solution properties [114, 115] of PFOA in supercritical CO_2 have been thoroughly investigated. In addition, we have shown that the lipophilic nature of the backbone of these polymers can be increased by incorporation of other monomers to make random copolymers [23].

Another approach to the stabilization of CO_2-phobic polymer colloids is the use of copolymerizable stabilizers. Rather than adsorbing to the surface of the growing polymer particle because of favorable enthalpic interactions, this type of lyophilic stabilizer actually chemically grafts to the surface of the particle. Examples of copolymerizable stabilizers include macroinitiators, macro-chain transfer agents, and macromonomers, all of which must contain a long CO_2-philic tail to be effective. The ability to use very small amounts of stabilizer constitutes one advantage of this type of approach [116]. With this technique, a block or graft copolymer which acts as a stabilizing moiety is generated in situ.

Finally, we have designed and synthesized a series of block copolymer surfactants for CO_2 applications . It was anticipated that these materials would self-assemble in a CO_2 continuous phase to form micelles with a CO_2-phobic core and a CO_2-philic corona. For example, fluorocarbon-hydrocarbon block copolymers of PFOA and PS were synthesized utilizing controlled free radical methods [104]. Small angle neutron scattering studies have demonstrated that block copolymers of this type do indeed self-assemble in solution to form multimolecular micelles [117]. Figure 5 depicts a schematic representation of the micelles formed by these amphiphilic diblock copolymers in CO_2. Another block copolymer which has proven useful in the stabilization of colloidal particles is the siloxane based stabilizer PS-*b*-PDMS [118, 119]. Chemical

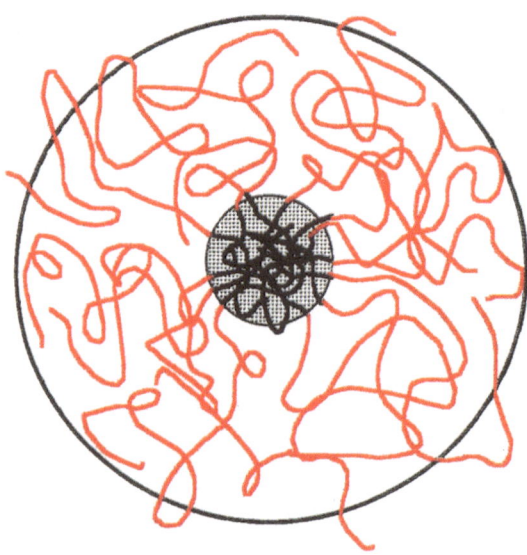

Fig. 5. Schematic of micelle formed by amphilic diblock copolymer

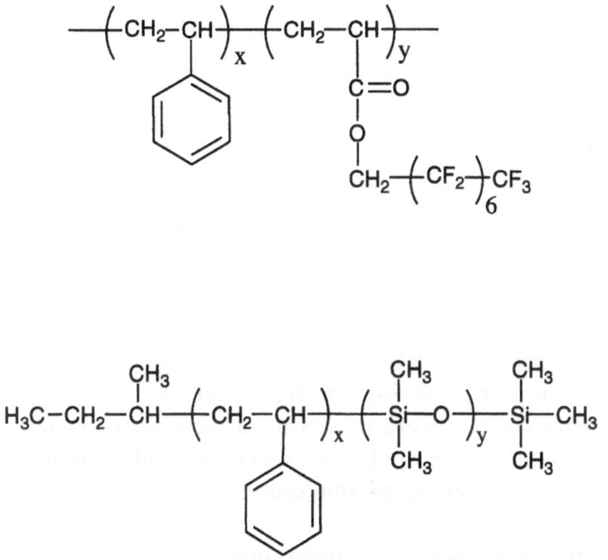

Fig. 6. Chemical structures of amphiphilic diblock copolymeric stabilizers PS-*b*-PFOA (*top*) and PS-*b*-PDMS (*bottom*) [118]

structures of these amphiphilic diblock copolymers are shown in Fig. 6. These types of surfactants which contain lipophilic segments that form the cores of the micelles in CO_2 may find use in a variety of applications ranging from polymer processing to precision cleaning and separations.

3.2.2 Polymerization of Lipophilic Monomers

In 1994, we reported the dispersion polymerization of MMA in supercritical CO_2 [103]. This work represents the first successful dispersion polymerization of a lipophilic monomer in a supercritical fluid continuous phase. In these experiments, we took advantage of the amphiphilic nature of the homopolymer PFOA to effect the polymerization of MMA to high conversions (>90%) and high degrees of polymerization (>3000) in supercritical CO_2. These polymerizations were conducted in CO_2 at 65 °C and 207 bar, and AIBN or a fluorinated derivative of AIBN were employed as the initiators. The results from the AIBN initiated polymerizations are shown in Table 3. The spherical polymer particles which resulted from these dispersion polymerizations were isolated by simply venting the CO_2 from the reaction mixture. Scanning electron microscopy showed that the product consisted of spheres in the μm size range with a narrow particle size distribution (see Fig. 7). In contrast, reactions which were performed in the absence of PFOA resulted in relatively low conversion and molar masses. Moreover, the polymer which resulted from these precipitation

Table 3. Results of MMA polymerization with AIBN as the initiator in CO_2 at 204 bar and 65 °C; stabilizer is either low molecular weight (LMW) or high molecular weight (HMW) PFOA [103]

Stabilizer (w/v %)	Yield (%)	$\langle M_n \rangle$ ($\times 10^{-3}$ g/mol)	PDI	Particle Size (μm)
0%	39	149	2.8	–
2% LMW	85	308	2.3	1.2 ($\pm$0.3)
4% LMW	92	220	2.6	1.3 ($\pm$0.4)
2% HMW	92	315	2.1	2.7 ($\pm$0.1)
4% HMW	95	321	2.2	2.5 ($\pm$0.2)

polymerizations had a random morphology which contrasted sharply to the spherical polymer particles which were produced in the dispersion polymerizations. Without a doubt, the amphiphilic macromolecule employed played a vital role in the stabilization of the growing PMMA colloidal particles.

Our recent work in this area has determined that very low amounts (0.24 wt %) of PFOA are needed to prepare a stable dispersion of PMMA latex particles [33, 120]. This work demonstrated that the stabilizer could be subsequently removed from the PMMA product by extraction with CO_2. Because of the relatively high cost of the stabilizer and the possible effects that residual stabilizer may have on product performance, the ability to remove and recycle the PFOA constitutes an important aspect of this system. In addition, the effect of the reaction time and pressure on the resulting conversion, molar masses, and particle size of the polymer product was investigated. The results from this work indicate that a gel effect occurs within the PMMA particles between one and two hours of reaction time. This result parallels the gel effect within the polymer particles which is normally observed between 20–80% conversion in a typical dispersion polymerization in liquid organic media [98]. More importantly, the ability of CO_2 to plasticize PMMA [4, 39–43] facilitates the diffusion of monomer into the growing polymer particles and causes autoacceleration which allows the reaction to proceed to high conversion. To complement the use of fluorinated acrylates as stabilizers, the phase behavior of PFOA in CO_2 was thoroughly investigated [33]. These cloud point experiments indicated lower critical solution temperature (LCST) phase behavior and confirmed the much higher solubility of fluorinated acrylate polymers in comparison to hydrocarbon polymers in CO_2.

The use of a copolymerizable macromonomer constitutes another approach to the dispersion polymerization of MMA. We have recently demonstrated the utility of this approach in CO_2 by employing a PDMS monomethacrylate as the stabilizer (see Fig. 8) [121]. Although several groups have studied the behavior of polysiloxanes in CO_2 [54, 55, 57], this work represents the first successful use of PDMS based polymeric stabilizers in CO_2. The reactions were conducted in either liquid CO_2 at 30 °C and 75 bar or supercritical CO_2 at 65 °C and 340 bar.

a)

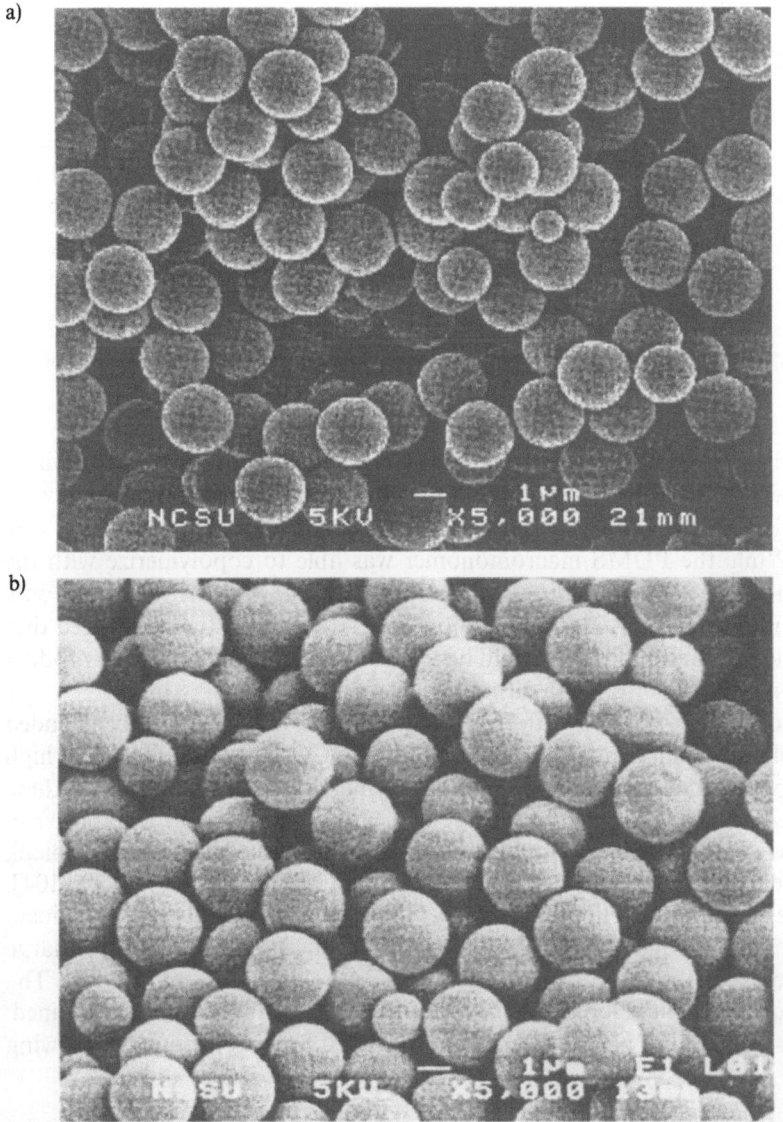

b)

Fig. 7. Scanning electron micrographs of PMMA particles produced by dispersion polymerization in supercritical CO_2. Stabilized by PFOA homopolymer (top) [103]; stabilized by PDMS macromonomer (bottom) [121]

Reactions were run at 20% (v/v) solids and azo initiators, which do not chain transfer to the PDMS backbone, were employed. Control reactions which were run in the absence of stabilizer or with PDMS homopolymer (which does not contain a polymerizable group) resulted in lower conversions and molecular

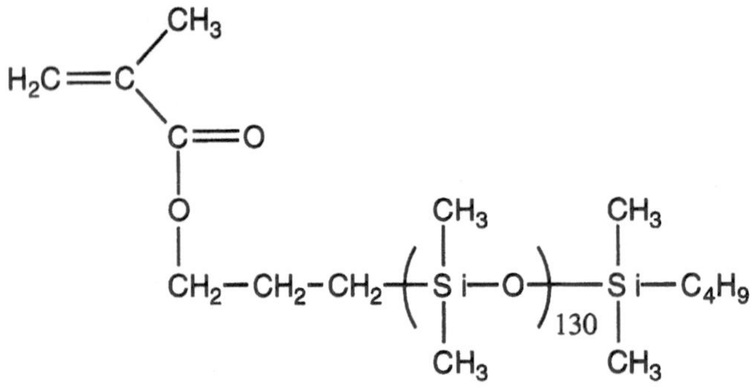

Fig. 8. Structure of PDMS macromonomer used as a copolymerizable stabilizer [121]

weights than the reactions which contained a small amount of the methacrylate-terminated PDMS macromonomer. Moreover, the polymerizations which were stabilized by the macromonomer resulted in spherical PMMA particles (see Fig. 7). While the PDMS macromonomer was able to copolymerize with the MMA to stabilize the growing polymer colloid, it was demonstrated that only a small amount of the PDMS macromonomer reacted with the MMA and that the unreacted macromonomer could be removed from the final polymer product by washing or extracting with either hexanes or CO_2.

In addition to our work with the PMMA system, we have recently extended the use of amphiphilic molecules as stabilizers in CO_2 to synthesize PS in high conversions (>95%) and high degrees of polymerization (>800) [122]. In these polymerizations, the first successful use of a block copolymer as a stabilizer in CO_2 was demonstrated. Initially, the stabilizers employed were diblock copolymers of PS (the anchoring block) and PFOA (the soluble block) [104]. Attempts to use PFOA homopolymer as the stabilizing moiety gave mediocre results, indicating that the CO_2-phobic nature of the PS block provides a large driving force for these copolymers to exist at the polymer-solvent interface. The results from these polymerizations are summarized in Table 4. The PS obtained from these dispersion polymerization was isolated as a dry, white, free-flowing

Table 4. Results of styrene polymerization with AIBN as the initiator in CO_2 at 204 bar and 65 °C; stabilizer is either PFOA homopolymer or PS-b-PFOA copolymer [122]

Stabilizer	Yield (%)	$\langle M_n \rangle$ ($\times 10^3$ g/mole)	PDI	Particle Diameter (μm)	Particle Size Distribution
none	22.1	3.8	2.3	none	none
poly(FOA)	43.5	12.8	2.8	none	none
(3.7 K/16 K)	72.1	19.2	3.6	0.40	8.3
(4.5 K/25 K)	97.7	22.5	3.1	0.24	1.3
(6.6 K/35 K)	93.6	23.4	3.0	0.24	1.1

powder directly from the reaction vessel. Scanning electron microscopy demonstrated that the product existed as submicron sized particles (see Fig. 9). In this
work, we showed that the diameters and size distributions of the resulting PS
particles could be controlled by variations in the ASB of the block copolymeric
stabilizer. Further studies focusing on the use of amphiphilic block copolymers
in a stabilizing capacity have employed the silicone based diblock copolymer
PS-b-PDMS [119]. This stabilizing system offers several advantages over the
fluorinated stabilizers, such as ease of synthesis and characterization, relatively
low cost, and narrow MWD of the block segments which results from the living
anionic techniques which are used to prepare these copolymers. This type of
silicone based block copolymer proved to be effective for the stabilization of
both styrene and MMA polymerizations in supercritical CO_2 at 65 °C and
345 bar using AIBN initiation. These experiments resulted in high yields
(>90%) of fine, dry white polymer powders. The PS particles which resulted
from these reactions possessed a M_n of 6.5×10^4 g/mol and an average particle
diameter of 0.22 μm; the PMMA particles had a M_n of 1.8×10^5 g/mol and an
average particle diameter of 0.23 μm. It should be noted that the PMMA
particles which are produced when the PS-b-PDMS is employed as the stabilizer
are about ten times smaller than those which are produced when PFOA
homopolymer is used [33]. In contrast, the analogous PDMS homopolymer
was not effective as a stabilizer in these systems, again indicating that the
presence of the PS anchoring group is vital to the interfacial activity of these
stabilizers.

3.2.3 Polymerization of Hydrophilic Monomers

Adamsky and Beckman have explored the possibility of carrying out an
inverse emulsion polymerization of acrylamide in supercritical CO_2 [123, 124].
In these reactions, acrylamide was polymerized in the presence of water,
a cosolvent for the monomer, in a CO_2 continuous phase at 345 bar and 60 °C;
AIBN was used as the initiator. Reactions were conducted both with and
without the stabilizer. The structure of the amide endcapped poly(hexafluoropropylene oxide) surfactant which was employed in these polymerizations
is shown in Fig. 10. The results from these polymerizations are shown in
Table 5. In the absence of the stabilizer, the precipitation polymerization
of acrylamide resulted in a high conversion of high molecular weight polymer
which formed a single solid mass in the bottom of the reaction vessel. In
the presence of stabilizer, the reaction solution had a milky white appearance
which was indicative of latex formation. No kinetic evidence or electron micrographs of colloidal particles were included in this article to support the
authors' hypothesis that the surfactant used was indeed effective in stabilizing
the polymer particles in an inverse emulsion polymerization. In fact, there
was no significant increase in the rate of the polymerization with added
surfactant.

a)

b)

c)

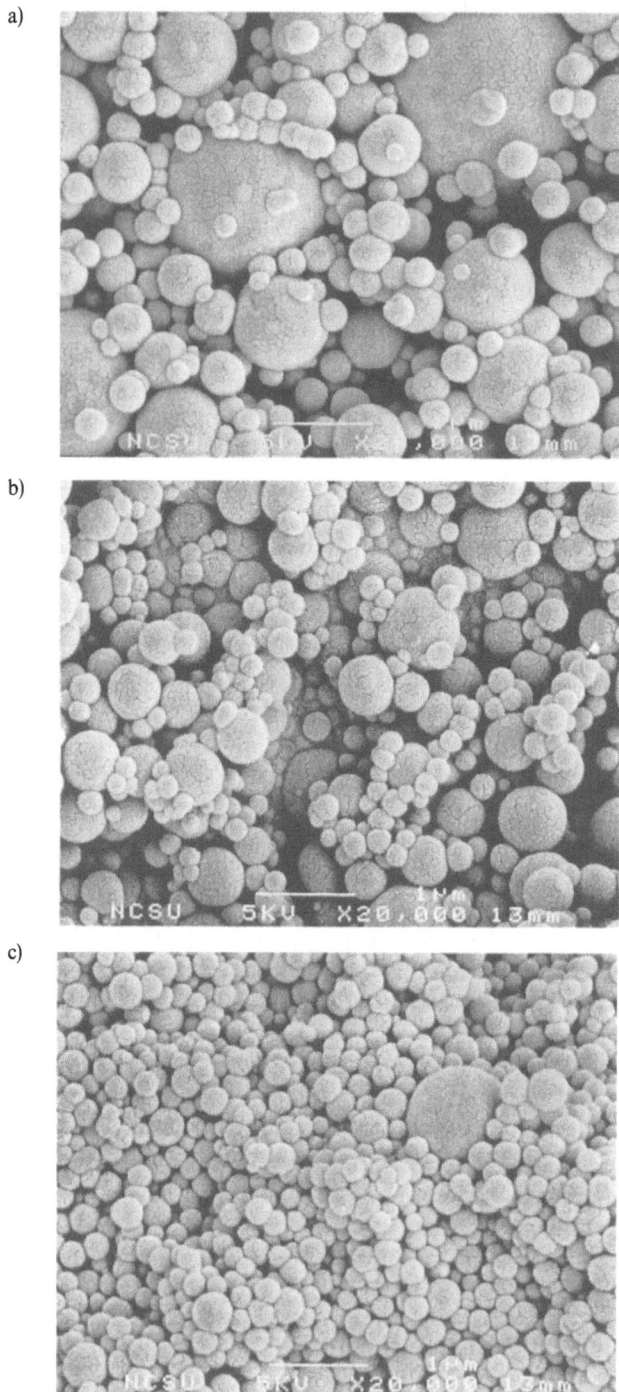

Fig. 9. Scanning electron micrographs of PS particles synthesized by dispersion polymerization utilizing **a** 3.6 K/16 K, **b** 4.5 K/25 K, and **c** 6.6 K/36 K PS-*b*-PFOA stabilizer (Mn PS segment/Mn

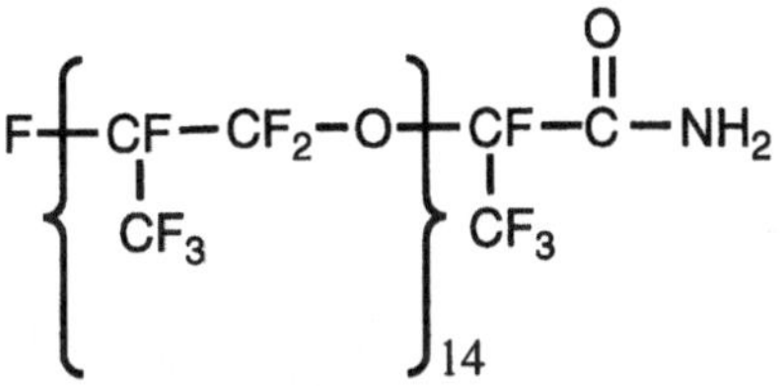

$$F\left\{CF-CF_2-O\right.\left\{\begin{array}{c}O\\\|\\CF-C-NH_2\\|\\CF_3\end{array}\right\}_{14}$$

Fig. 10. Amide end-capped poly(hexafluoropropylene oxide) used for the inverse emulsion polymerization of acrylamide in CO_2 [123]

Table 5. Results of acrylamide polymerization in CO_2 at 60 °C [124]

Polymer Property	Concentration of fluorinated polyether		
	0%	1%	2%
intrinsic viscosity (dL/g)	11.60	12.28	9.15
$M_v \times 10^{-6}$	6.61	7.09	4.92
Huggins constant	0.310	0.505	0.479
percent yield (wt %)	91.4	99.8	99.8

3.2.4 Polymer Blends

Watson and McCarthy have extended the idea of polymerization being facilitated by plasticization of the polymer phase to develop a new route to polymer blends [31, 32, 125]. In their system, a supercritical CO_2 swollen solid polymer substrate is infused with styrene monomer which is subsequently polymerized in situ (see Fig. 11). In these experiments, the solid polymer matrices explored include poly(chlorotrifluoroethylene) (PCTFE), poly(4-methyl-1-pentene) (PMP), high density polyethylene (HDPE), nylon-6,6-poly(oxymethylene), and bisphenol A polycarbonate. In these experiments, either AIBN or *tert*-butyl perbenzoate were used as the initiator for the polymerizatioin of the styrene within the supercritical fluid swollen matrix polymer. The authors employed transmission electron microscopy and energy dispersive X-ray analysis to demonstrate that the polystyrene exists as discrete phase-segregated regions throughout the matrix polymer (PCTFE in this case). They also used thermal analysis to show that radical grafting reactions are not significant. Indeed, exploitation of the enhanced diffusion of small molecules into polymeric phases which are plasticized by CO_2 opens up an entirely new region of research.

3.3 Heterogeneous Cationic Polymerizations

Cationic polymerization represents another area in which CO_2 has proven to be a viable continuous phase. In this area, both the liquid and supercritical phases

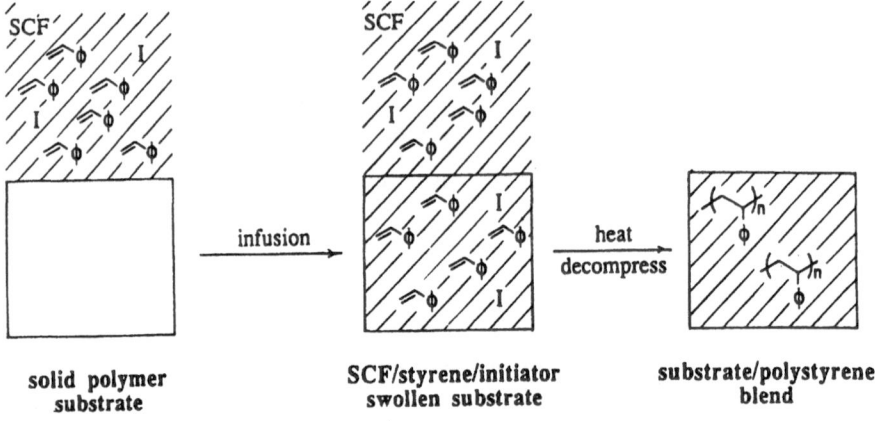

solid polymer   SCF/styrene/initiator   substrate/polystyrene
   substrate   swollen substrate   blend

Fig. 11. Schematic of new route to composite polymer materials [31, 32]

have been investigated. One of the interesting aspects of using the supercritical phase in particular may lie in the ability to change the dielectric constant of the solvent by simply varying the temperature or pressure and hence change the intimacy of the ion pair in the propagating species. This type of work could have tremendous implications in controlling the "livingness" of these cationic system, which are often plagued by chain transfer to monomer. Preliminary work which has been conducted in the area of heterogeneous cationic polymerizations in CO_2 has focused on monomers such as isobutylene, formaldehyde, vinyl ethers, and oxetanes.

The most industrially significant polymerizations involving the cationic chain growth mechanism are the various polymerizations and copolymerizations of isobutylene. In fact, about 500 million pounds of butyl rubber, a copolymer of isobutylene with small amounts of isoprene, are produced annually in the United States via cationic polymerization [126]. The necessity of using toxic chlorinated hydrocarbon solvents such as dichloromethane or methyl chloride as well as the need to conduct these polymerizations at very low temperatures constitute two major drawbacks to the current industrial method for polymerizing isobutylene which may be solved through the use of CO_2 as the continuous phase.

In 1960, Biddulph and Plesch reported the first example of synthesis of polyisobutylene in liquid CO_2 at $-50\,°C$ using titanium tetrachloride or aluminum bromide as the catalyst [127]. Recently, Pernecker and Kennedy have been exploring the polymerization of isobutylene in CO_2 at supercritical conditions with added cosolvents [128–132]. In this work, the authors demonstrated that polyisobutylene could be synthesized in CO_2 with methyl chloride at $32.5\,°C$ and approximately 120 bar using 2-chloro-2,4,4-trimethylpentane (TMPCl)/ $SnCl_2$ and TMPCl/TiCl$_4$ initiating systems. These reactions typically gave conversions ranging from 7–35% of polymer products with M_ns in the range 1000–2000 g/mol and PDIs from 1.5 to 3.4 [128]. In this work, the authors note that this represents the highest temperature at which isobutylene has even been polymerized to achieve a reasonably high molecular weight polymer. They attributed this ability to form a polymer at high temperatures to the nature of the supercritical solvent, and point out that this process would save considerable expense in industry where cationic polymerization systems must presently be cooled to reaction temperatures in the $-20\,°C$ to $-100\,°C$ range to minimize the chain transfer reactions inherent to cationic systems which limit the molecular weights of the resulting polymers [131]. Other exploratory work by these authors include the synthesis of *tert*-Cl terminated polyisobutylene by mixed Friedel-Crafts acid initiating systems [129], the determination of the ceiling temperature of isobutylene polymerization in supercritical CO_2 [130], and the synthesis of poly(isobutylene-*co*-styrene) [131].

In the late 1960s, the cationic polymerization of formaldehyde in liquid carbon dioxide was investigated by Fukui and coworkers [133–137]. In this work, polymerizations were conducted in liquid CO_2 at temperatures ranging from 0–50 °C and pressures ranging from 15 to 34 bar. The reactions resulted in white polymer products which had yields as high as 70% and degrees of polymerization as high as 2200. Even without the addition of any catalyst, the acidic impurities in the monomer were sufficient to initiate the reaction to give polymers of high degrees of polymerizations. Addition of an acidic catalyst such as a carboxylic acid increased both the yield and molecular weight of the resulting polymer. Furthermore, the authors discovered that the chain transfer to impurities was suppressed in these reactions, so that even relatively impure monomer could be polymerized to give a high molecular weight product. The motivation of this work was that in conventional polymerizations the liquid formaldehyde must be rigorously purified to remove even trace amounts of impurities which can act as chain transfer agents and destroy the resproducibility of the results. The authors were following up on their discovery that small amounts of CO_2 noticeably depress the polymerization [138] and they were able to use a CO_2 continuous phase to control the reaction and obtain reproducible results [133]. They conducted a kinetic study and determined that the overall rate of the reaction was first order with respect to the monomer [134]. They also investigated the role of various monomer preparation methods [135] and developed detailed mechanisms for the elementary processes in the polymerization [136]. In 1969 the authors received a United States patent [137] for this process.

Vinyl ethers constitute a third class of monomers which have been cationically polymerized in CO_2. While fluorinated vinyl ether monomers such as those described in Sect. 2.1.2 can be polymerized homogeneously in CO_2 because of the high solubility of the resulting amorphous fluoropolymers, the polymerization of hydrocarbon vinyl ethers in CO_2 results in the formation of CO_2-insoluble polymers which precipitate from the reaction medium. The work in this area reported to date in the literature includes precipitation polymerizations and does not yet include the use of stabilizing moieties such as those described in the earlier sections on dispersion and emulsion polymerizations (Sect. 3).

In 1970, a patent by Fukui and coworkers cited the cationic polymerization of ethyl vinyl ether in liquid CO_2 [85]. In these reactions, $SnCl_4$ or ethyl etherate of boron trifluoride were employed as the catalysts and the polymerizations were conducted for 20 hours at room temperature to conversion of greater than 90% polymer. No molecular weight data or spectra for these polymers were reported in this work.

Recently, we have investigated the polymerization of isobutyl vinyl ether (IBVE) in supercritical CO_2 at 30–60 °C and 345 bar using a known initiating system consisting of the adduct of acetic acid and IBVE as the initiator, the Lewis acid ethyl aluminum dichloride as the coinitiator, and ethyl acetate as the Lewis base deactivator [78, 139]. This initiating system was previously developed by Higashimura for the living cationic polymerization of vinyl ethers at higher temperatures in traditional liquid solvents [140]. Although the polymerizations became heterogeneous as the insoluble hydrocarbon polymer precipitated from the CO_2 solution, high molecular weights and high conversions were obtained. Control experiments performed in cyclohexane gave comparable conversion and M_n results but had lower molecular weight distributions.

Because a review of previous work revealed that CO_2 could act as a monomer in some polymerizations catalyzed by Lewis acids [141], for our systems it was important to demonstrate that the supercritical CO_2 being employed as the continuous phase was not being incorporated into the backbone of the polymer chain. Spectral analysis consisting of ^1H and ^{13}C NMR as well as infrared spectroscopy demonstrated that no differences existed in the structure of the polymers prepared in hexane and those prepared in CO_2, proving that the CO_2 was acting as an inert solvent in these polymerizations and was not acting as a monomer [139].

Strained cyclic ethers such as epoxides and oxetanes constitute an additional class of monomers which have been polymerized via heterogeneous cationic techniques in CO_2. Early work in this area by the Sumitomo Chemical Company and Seitetso Kagaku Company focused on the ring opening polymerization of epoxides such as ethylene oxide, propylene oxide, and styrene oxide [142]. These polymerizations employed catalyst systems such as triethyl aluminum, aluminum bromide, diethyltin/water, and titanium tetrachloride and were conducted at 60 °C. In general, the reported reactions resulted in powdery polymer products which were recovered in yields ranging from 32% to 93%. The reduced viscosities of the polymers formed were between 0.6 and 2.7. The authors pointed out both the ease of recovering the product, such as PEO, which was isolated dry after venting to remove the CO_2 and the ability to use supercritical fluid extraction methods to remove residual monomer. In more current work, we have explored the polymerization of cyclic ethers (oxetanes) in liquid CO_2 [78]. In contrast to the homogeneous polymerization of the fluorinated oxetane FOx-7 (see Sect. 2.1.2), the polymerization of 3,3'-bis(ethoxymethyl)oxetane (BEMO) in CO_2 at −10 °C and 290 bar for four hours using BF_3 as the initiator results in the precipitation of the polymer product as it forms. It was demonstrated that control solution polymerizations of BEMO which were conducted in dichloromethane resulted in yields ($\sim 70\%$) and molar masses ($M_n \sim 10^4$ g/mol) which were similar to those resulting from the heterogeneous polymerizations conducted in CO_2. Again, structural determination of the polymer products from the reactions conducted in CO_2 showed that the CO_2 was not incorporated into the backbone of the polymer chain.

Finally, researchers at Rhone-Poulenc Specialty Chemicals field a European patent application for the preparation of polydiorganosiloxanes in supercritical fluids in 1986 [143]. Several example reactions cited in this work illustrate the polymerization of octamethylcyclotetrasiloxane (D_4) in CO_2. In these reactions, trifluoromethane sulfonic acid was used as the initiator; the authors noted that with an acidic fluid such as CO_2 one cannot use a basic catalyst. These polymerizations were conducted in a 100 mL high pressure reactor with initial concentrations of 2.0 mol L^{-1} of D_4 and 6.0×10^{-3} mol L^{-1} of initiator in three different physical states of carbon dioxide: gas, liquid, and supercritical fluid. The polymerizations conducted in the gaseous CO_2 resulted in the highest yields and molecular weights of poly(dimethylsiloxane) (PDMS). This result is not surprising, since these reactions were simply bulk reactions under an atmosphere of CO_2. A series of reactions were conducted in supercritical CO_2 at pressures ranging from 8–190 bar and temperatures ranging from 52–82 °C with reaction times of up to one hour. The resulting gravimetric yields from these reactions were from 5–98% with M_n ranging from 8.3×10^4 g/mol to 2.2×10^5 g/mol. Again, this type of polymerization illustrates the viability of performing cationic polymerizations in a CO_2 medium.

3.4 Metal Catalyzed Polymerizations

3.4.1 Ring Opening Metathesis Polymerizations

We have reported the first example of a ring-opening metathesis polymerization in CO_2 [144, 145]. In this work, bicyclo[2.2.1]hept-2-ene (norbornene) was polymerized in CO_2 and CO_2/methanol mixtures using a $Ru(H_2O)_6(tos)_2$ initiator (see Scheme 6). These reactions were carried out at 65 °C and pressure was varied from 60 to 345 bar; they resulted in poly(norbornene) with similar conversions and molecular weights as those obtained in other solvent systems. ^1H NMR spectroscopy of the poly(norbornene) showed that the product from a polymerization in pure methanol had the same structure as the product from the polymerization in pure CO_2. More interestingly, it was shown that the cis/trans ratio of the polymer microstructure can be controlled by the addition of a methanol cosolvent to the polymerization medium (see Fig. 12). The poly(norbornene) prepared in pure methanol or in methanol/CO_2 mixtures had a very high *trans*-vinylene content, while the polymer prepared in pure CO_2 had very high *cis*-vinylene content. These results can be explained by the solvent effects on relative populations of the two different possible metal

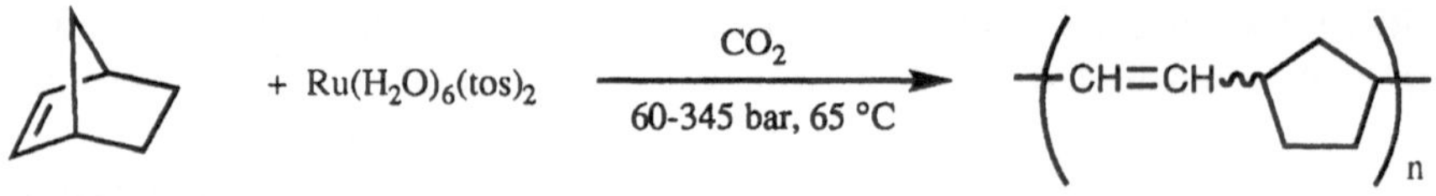

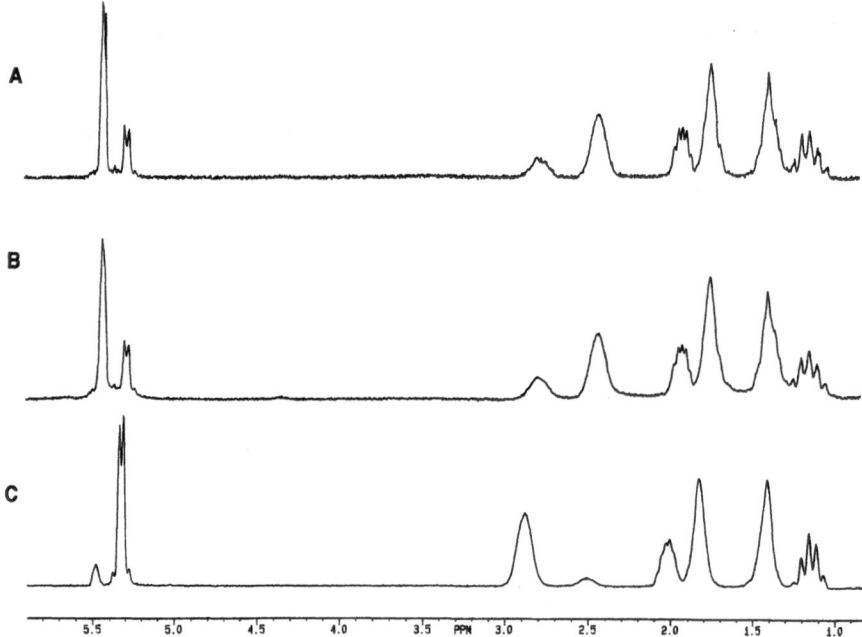

Scheme 6. Ring-opening metathesis polymerization of norbornene in CO_2 [144, 145]

Fig. 12. The ^1H NMR of poly(norbornene) prepared in **A** methanol **B** methanol/CO_2 and **C** CO_2 [144]

carbene propagating species which give rise to the different microstructures. Investigations involving the use of other organometallic initiating systems and other cyclic olefin monomers are underway. Indeed, the expansion of polymer synthesis in CO_2 to include metal catalyzed reaction mechanisms represents an important step in the use of CO_2 as a universal solvent for polymerizations.

3.4.2 Oxidative Coupling Polymerizations

We have recently demonstrated the synthesis of poly(2,6-dimethylphenylene oxide) (PPO) via oxidative coupling in a CO_2 continuous phase (see Scheme 7) [76]. These reactions proceeded at 345 bar and either room temperature or

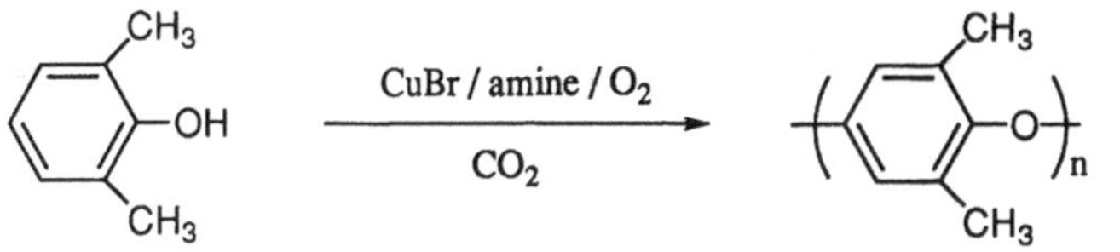

Scheme 7. Oxidative coupling polymerization of 2,6-dimethylphenol in CO_2 [76]

40 °C for 20 hours. A catalyst system based on copper bromide, an amine, and oxygen was employed in these polymerizations, resulting in PPO yields as high as 83% and M'_ns as high as 1.7×10^4 g/mol.

Parameters which were varied in the system include the structure of the amine and the addition of polymeric stabilizers. To begin with, the effectiveness of both traditional small molecule amines (pyridine and dimethylethylamine) and polymeric, CO_2-soluble amines (random copolymers of FOA and either 4-vinylpyridine or 2-((dimethylamino)ethylacrylate) was examined. While the results from this study indicated that pyridine is the most effective amine, the polymerizations conducted in the presence of the polymeric amines resulted in a milky white solution characteristic of a latex, indicating that these polymeric amines also served to stabilize the PPO as a polymer colloid. A second variable which played a key role in these oxidative coupling polymerizations was the addition of a stabilizing moiety. To this end, the homopolymer PFOA, the block copolymer PS-*b*-PFOA, and the random copolymer PS-*co*-PFOA were assessed in their ability to improve the resulting yield and molecular weight of PPO. Because CO_2 plasticizes PPO [47] and facilitates the transport of monomer into the polymer phase, stabilizing the PPO as a colloidal dispersion should simultaneously lead to an increase in both of these properties. The block copolymer was shown to be the most effective stabilizer, and the stabilized reaction did indeed result in both higher yields and molecular weights than the precipitation polymerization in the absence of stabilizer.

3.5 Step Growth Polymerizations

Many of examples cited in this text and elsewhere [7] illustrate that chain growth polymerization mechanisms have been widely studied in compressible fluids such as CO_2. Step growth polymerizations, on the other hand, constitute an emerging area of study in these types of fluids. We have begun investigating this type of polymerizations through the synthesis of aromatic polyesters such as PET in supercritical CO_2 [146]. The strategy underlying our approach to this challenge involves removal of the small molecule byproduct of transesterification by supercritical fluid extraction. In this approach, the removal of the condensate molecule, ethylene glycol, is facilitated by the ability of CO_2 to act as a plasticizing agent for the polymer phase (see Scheme 8). This method

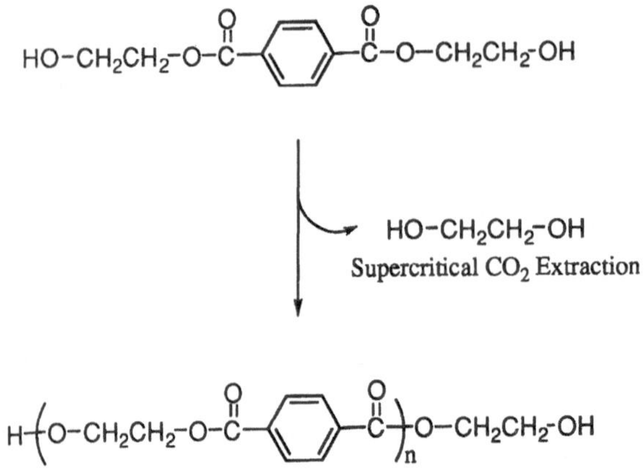

HO−CH₂CH₂−O−C(=O)−⟨benzene ring⟩−C(=O)−O−CH₂CH₂−OH

HO−CH₂CH₂−OH
Supercritical CO_2 Extraction

H−(O−CH₂CH₂−O−C(=O)−⟨benzene ring⟩−C(=O)−O−CH₂CH₂−OH)ₙ

Scheme 8. Synthesis of PET by supercritical CO_2 extraction method [146]

provides an alternative to the current industrial process which requires very high temperatures and vacuum to allow the removal of the condensate molecule from the polymer melt. The enhanced removal of the ethylene glycol in the CO_2-based process may result in increased rates of polymerization at lower temperatures than those required for the conventional methods in the absence of CO_2. Additionally, when combined with the recent advances in surfactant design, this technique for the removal of the small molecule condensate by supercritical CO_2 extraction may allow for the extension of this general strategy to the synthesis of other industrially important classes of polymers which are made by a step growth mechanism.

3.6 Hybrid Systems

We have developed a hybrid polymerization methodology based on two phase mixtures of CO_2 and water for the synthesis of high molar mass polymers such as PTFE [147, 148]. This allows for the compartmentalization of monomer, polymer, and initiator based on their solubility characteristics. Another advantage of this system is that highly exothermic polymerizations can be somewhat controlled by the high heat capacity of the water phase. In the newly developed hybrid system, water soluble persulfate initiators were used at 75 °C and sodium perfluorooctanoate was used as the surfactant. Using this process, reasonably high molar mass ($M_n = 1 \times 10^6$ g/mol) PTFE was produced in good yields (80–90%). Other experiments at ambient temperature which employed a redox initiator (sodium sulfite and iron (II) salt were added as reducing agents) gave similar results. Indeed, the combination of the solvent properties of water and CO_2 may prove to be advantageous in a variety of systems, including the synthesis of structured latex composite particles due to the plasticization of the growing polymer particles by added CO_2.

4 Conclusions

Recent progress has demonstrated that CO_2 represents a viable and attractive polymerization medium. For free radical reactions, numerous studies have shown that CO_2 may prove to be the solvent of choice for many applications. In addition, other chain growth mechanisms for polymer propagation such as cationic and ring opening metathesis polymerizations in CO_2 have been explored and advantages of this solvent have been demonstrated. Moreover, the most recent work has shown that it also represents a good solvent in which to conduct step-growth polymerizations. Specially designed stabilizers for use in CO_2 constitutes another area which will allow the extension of the use of CO_2 as a continuous phase. The ability to tailor the surfactant molecule for a particular CO_2-based application may prove to be the enabling technology for the reduction of the use of hazardous VOCs and CFCs. When the environmental advantages of CO_2 are combined with its ability to be used as a solvent for a wide variety of chemical reactions, one can make compelling arguments that CO_2 may be the solvent of the future for the polymer industry.

Acknowledgements. We gratefully acknowledge financial support from the National Science Foundation through a Presidential Faculty Fellowship (JMD: 1993–1997), the Environmentally Benign Chemical Synthesis and Processing Program sponsored by NSF and the Environmental Protection Agency, and the *Consortium for Polymeric Materials Synthesis and Processing in Carbon Dioxide* sponsored by DuPont, Air Products and Chemicals, Hoechst-Celanese, Eastman Chemical, B.F. Goodrich, Xerox, Bayer, and General Electric.

5 References

1. Johnston KP, Penninger JML (1989) Supercritical fluid science and technology. In: Comstock MJ (ed) ACS Symposium Series, vol 406. ACS, Washington
2. Bruno TJ, Ely JF (1991) Supercritical fluid technology: Reviews in Modern Theory and Applications. CRC Press, Boston
3. Kiran E, Brennecke JF (1993) Supercritical fluid engineering science: Fundamentals and applications. In: Comstock MJ (ed) ACS Symposium Series, vol 514, ACS, Washington
4. McHugh MA, Krukonis VJ (1993) Supercritical fluids extraction: Principles and practice. Butterworth-Heineman, Stoneham
5. Savage PE, Gopalan S, Mizan TI, Martino CJ, Brock EE (1995) AIChE J 41: 1723
6. Scholsky KM (1993) J Supercrit Fluids 6: 103
7. Shaffer KA, DeSimone JM (1995) Trends Polym Sci 3: 146
8. Barer SJ, Stern KM (1988) Sources and economics of carbon dioxide. In: Ayers WM (ed) Catalytic activation of carbon dioxide. American Chemical Society, Washington, D.C. p 1
9. Quinn EL, Jones CL (1936) Carbon Dioxide. Reinhold, New York
10. Kaupp G (1994) Angew Chem Int Ed Engl 33: 1452
11. Jessop PG, Ikariya T, Noyori R (1995) Organometallics 14: 1510
12. Burk MJ, Feng S, Gross MF, Tumas W (1995) J Am Chem Soc 117: 8277
13. Keyes FG, Kirkwood JG (1930) Phys Rev 36: 754
14. McFann GJ, Johnston KP, Howdle SM (1994) AIChE J 40: 543

15. Hyatt JA (1984) J Org Chem 49: 5097
16. O'Shea KE, Kirmse WR, Fox MA, Johnston KP (1991) J Phys Chem 95: 7863
17. Lowry HH, Erickson WR (1927) J Am Chem Soc 49: 2729
18. King MB, Mubarak A, Kim JD, Bott TR (1992) J Supercrit Fluids 5: 296
19. Bartle KD, Clifford AA, Jafar SA, Shilstone GF (1991) J Phys Chem Ref Data 20: 728
20. Rindfleisch F, DiNoia T, McHugh MA (1996) Polym Mater Sci Eng 74: 178
21. Yilgor I, McGrath JE, Krukonis VJ (1984) Polym Bull 12: 499
22. Krukonis V (1985) Polymer News 11: 7
23. DeSimone JM, Guan Z, Elsbernd CS (1992) Science 257: 945
24. Guan Z, Combes JR, Menceloglu YZ, DeSimone JM (1993) Macromolecules 26: 2663
25. Hoefling TA, Newman DA, Enick RM, Beckman EJ (1993) J Supercrit Fluids 6: 165
26. Pilato LA, Litz LM, Hargitay B, Osborne RC, Farnham AG, Kawakami JH, Fritze PE, McGrath JE (1975) Polym Prepr (Am Chem Soc, Div Polym Chem) 16: 41
27. Fried JR, Li W (1990) J Appl Polym Sci 41: 1123
28. Shah VM, Hardy BJ, Stern SA (1993) J Polym Sci, Part B: Polym Phys 31: 313
29. Shah VM, Hardy BJ, Stern SA (1986) J Polym Sci, Part B: Polym Phys 24: 2033
30. Kazarian SG, Vincent MF, Bright FV, Liotta CL, Eckert CA (1996) J Am Chem Soc 118: 1729
31. Watkins JJ, McCarthy TJ (1994) Macromolecules 27: 4845
32. Watkins JJ, McCarthy TJ (1995) Macromolecules 28: 4067
33. Hsiao Y-L, Maury EE, DeSimone JM, Mawson SM, Johnston KP (1995) Macromolecules 28: 8159
34. Sand ML (1986) U.S. Pat 4,598,006
35. Berens AR, Huvard GS, Korsmeyer RW (1989) U.S. Pat 4,820,752
36. Berens AR, Huvard GS, Korsmeyer RW, Kunig FW (1992) J Appl Polym Sci 46: 231
37. Watkins JJ, McCarthy TJ (1995) Polym Mater Sci Eng 73: 158
38. Wang W-CV, Kramer EJ, Sachse WH (1982) J Polym Sci: Polym Phys Ed 20: 1371
39. Chiou JS, Barlow JW, Paul DR (1985) J Appl Polym Sci 30: 2633
40. Wissinger RG, Paulaitis (1987) J Polym Sci, Part B: Polym Phys 25: 2497
41. Wissinger RG, Paulaitis ME (1991) J Polym Sci, Part B: Polym Phys 29: 631
42. Goel SK, Beckman EJ (1993) Polymer 34: 1410
43. Condo PD, Paul DR, Johnston KP (1994) Macromolecules 27: 365
44. Condo PD, Johnston KP (1994) J Polym Sci, Part B: Polym Phys 32: 523
45. Kamiya Y, Hirose T, Mizoguchi K, Terada K (1988) J Polym Sci, Part B: Polym Phys 26: 1409
46. Kamiya Y, Mizoguchi K, Hirose T, Naito Y (1989) J Polym Sci, Part B: Polym Phys 27: 879
47. Handa YP, Lampron S, O'Neill ML (1994) J Polym Sci, Part B: Polym Phys 32: 2549
48. Briscoe BJ, Kelly CT (1995) Polymer 36: 3099
49. Shieh Y-T, Su J-H, Manivannan G, Lee PHC, Sawan SP, Spall WD (1996) J Appl Polym Sci 59: 695
50. Shieh Y-T, Su J-H, Manivannan G, Lee PHC, Sawan SP, Spall WD (1996) J Appl Polym Sci 59: 707
51. Condo PD, Sanchez IC, Panayiotou CG, Johnston KP (1992) Macromolecules 25: 6119
52. Kalospiros NS, Paulaitis ME (1994) Chem Eng Sci 49: 659
53. McHugh MA, Krukonis VJ, Pratt JA (1994) Trends Polym Sci 2: 301
54. Elsbernd CS, DeSimone JM, Hellstern AM, Smith SD, Gallagher PM, Krukonis VJ, McGrath JE (1990) Polym Prepr (Am Chem Soc, Div Polym Chem) 31: 673
55. Zhao X, Watkins R, Barton SW (1995) J Appl Polym Sci 55: 773
56. Garg A, Gulari E, Manke CW (1994) Macromolecules 27: 5643
57. Mertsch R, Wolf BA (1994) Macromolecules 27: 3289
58. Kwag C, Gerhardt LJ, Khan V, Gulari E, Manke CW (1996) Polym Mater Sci Eng 74: 183
59. Elliott J, Jarrell R, Srinivasan G, Dhanuka M, Akhaury R (1992) U.S. Pat 5,128,382
60. Srinivasan G, Elliot J, Richard J (1992) Ind Eng Chem Res 31: 1414
61. Wessling M, Borneman Z, Van Den Boomgaard T, Smolders CA (1994) J Appl Polym Sci 53: 1497
62. Goel SK, Beckman EJ (1995) AIChE J 41: 357
63. Randolph TW, Clark DS, Blanch HW, Prausnitz JM (1988) Science 238: 387
64. Ikushima Y, Saito N, Arai M, Blanch HW (1995) J Phys Chem 99: 8941
65. Makin EC (1984) U.S. Pat 4,474,994
66. Hybertson BM, Repine JE, Beehler CJ, Rutledge KS, Lagalante AF, Sievers RE (1993) J Aerosol Med 6: 275

67. Van Bramer DJ, Shiflett MB, Yokozeki A (1994) U.S. Pat 5,345,013
68. Zosel K (1974) U.S. Pat 3,806,619
69. Prasad R, Gottesman M, Scarella RA (1982) U.S. Pat 4,341,804
70. Leyendecker D (1988) Selection of conditions for an SFC separation. In: Smith RM (ed) Supercritical fluid chromatography. Royal Society of Chemists, London p 53
71. Schwarzbach SE (1995) Nature 376: 297
72. Guan Z, Combes JR, Elsbernd CS, DeSimone JM (1993) Polym Prepr (Am Chem Soc, Div Polym Chem) 34: 446
73. Ehrlich P (1993) Chemtracts: Org Chem 6: 92
74. DeSimone JM (1993) Int Pat Appl PCT/US93/01626
75. Guan Z, Elsbernd CS, DeSimone JM (1992) Polym Prepr (Am Chem Soc, Div Polym Chem) 34: 329
76. Kapellen KK, Mistele CD, DeSimone JM (1996) Polym Mater Sci Eng 74: 256
77. Romack TJ, Combes JR, DeSimone JM (1995) Macromolecules 28: 1724
78. Clark MR, DeSimone JM (1995) Macromolecules 28: 3002
79. Hagiwara M, Mitsui H, Machi S, Kagiya T (1968) J Polym Sci: Part A-1 6: 603
80. Hagiwara M, Mitsui H, Machi S, Kagiya T (1968) J Polym Sci: Part A-1 6: 609
81. Hagiwara M, Mitsui H, Machi S, Kagiya T (1968) J Polym Sci: Part A-1 6: 721
82. Hagiwara M, Okamoto H, Kagiya T (1970) Bull Chem Soc Jap 43: 172
83. Kagiya T, Machi S, Hagiwara M, Kise S (1969) U.S. Pat 3,471,463
84. Sumitomo (1968) French Pat 1,524,533
85. Fukui K, Kagiya T, Yokota H, Toriuchi Y, Kuniyoshi F (1970) U.S. Pat 3,522,228
86. Hartmann H, Denzinger W (1986) Canadian Pat 1,262,995
87. Terry RE, Zaid A, Angelos C, Whitman DL (1988) Energy Prog 8: 48
88. Sertage WG, Jr., Davis P, Schenck HU, Denzinger W, Hartmann H (1986) Canadian Pat 1,274,942
89. Herbert MW, Huvard GS (1989) Eur Pat Appl 0,301,532
90. Romack TJ, Maury EE, DeSimone JM (1995) Macromolecules 28: 912
91. Dada E, Lau W, Merritt RF, H. PY, Swift G (1996) Polym Mater Sci Eng 74: 427
92. Romack TJ, DeSimone JM, Treat TA (1995) Macromolecules 28: 8429
93. Romack TJ, DeSimone JM (1996) Polym Mater Sci Eng 74: 428
94. Odell PG, Hamer GK (1996) Polym Mater Sci Eng 74: 404
95. Arshady R (1992) Colloid Polym Sci 270: 717
96. Smith WV, Ewart RH (1948) J Chem Phys 16: 592
97. Napper DH (1983) Polymeric stabilization of colloidal dispersions. Academic Press, New York
98. Barrett KEJ (1975) Dispersion polymerization in organic media. John Wiley, New York
99. Piirma I (1992) Polymeric Surfactants. Dekker, New York
100. Peck DG, Johnston KP (1993) Macromolecules 26: 1537
101. Peck DG, Johnston KP (1993) J Phys Chem 97: 5661
102. DeSimone JM, Maury EE, Combes JR, Menceloglu YZ (1993) Polym Mater Sci Eng 68: 41
103. DeSimone JM, Maury EE, Menceloglu YZ, McClain JB, Romack TR, Combes JR (1994) Science 265: 356
104. Guan Z, DeSimone JM (1994) Macromolecules 27: 5527
105. Consani KA, Smith RD (1990) J Supercrit Fluids 3: 51
106. Maury EE, Batten HJ, Killian SK, Menceloglu YZ, Combes JR, DeSimone JM (1993) Polym Prepr (Am Chem Soc, Div Polym Chem) 34: 664
107. Fulton JL, Pfund DM, McClain JB, Romack TR, Maury EE, Combes JR, Samulski ET, DeSimone JM, Capel M (1995) Langmuir 11: 4241
108. Newman DA, Hoefling TA, Beitle RR, Beckman EJ, Enick RM (1993) J Supercrit Fluids 6: 205
109. Harrison K, Goveas J, Johnston KP, O'Rear IEA (1994) Langmuir 10: 3536
110. Johnston KP, Harrison KL, Clarke MJ, Howdle SM, Heitz MP, Bright FV, Carlier C, Randolph TW (1996) Science 271: 624
111. Bartscherer KA, Minier M, Renon H (1995) Fluid Phase Equilibria 107: 93
112. McFann GJ, Johnston KP (1996) Supercritical Microemulsions. In: Kumar P, Mittal KL (eds) Microemulsions: Fundamentals and Applied Aspects, in press
113. Cohen-Stuart MA, Waajen FHWH, Cosgrove T, Vincent B, Crowley TL (1984) Macromolecules 17: 1825
114. McClain JB, Londono D, Combes JR, Romack TJ, Canelas DA, Betts DE, Wignall GD, Samulski ET, DeSimone JM (1996) J Am Chem Soc 118: 917

115. McClain JB, Betts DE, Canelas DA, Samulski ET, DeSimone JM, Londono JD, Wignall GD (1996) Polym Mater Sci Eng 74: 234
116. Kobayashi S, Uyama H, Lee SW, Matsumoto Y (1993) J Polym Sci, Part A: Polym Chem 31: 3133
117. Chillura-Martino D, Triolo R, McClain JB, Combes JR, Betts DE, Canelas DA, Samulski ET, DeSimone JM, Cochran HD, Londono JD, Wignall GD (1996) J Molec Struct: in press
118. DeSimone JM, Maury EE, Combes JR, Menceloglu YZ (1995) U.S. Pat 5,382,623
119. Canelas DA, Betts DE, DeSimone JM (1996) Polym Mater Sci Eng 74: 400
120. Hsiao Y-L, Maury EE, DeSimone JM (1995) Polym Prepr (Am Chem Soc, Div Polym Chem) 36: 190
121. Shaffer KA, Jones TA, Canelas DA, DeSimone JM, Wilkinson SP (1996) Macromolecules 29: 2704
122. Canelas DA, Betts DE, DeSimone JM (1996) Macromolecules 29: 2818
123. Adamsky FA, Beckman EJ (1994) Macromolecules 27: 312
124. Adamsky FA, Beckman EJ (1994) Macromolecules 27: 5238
125. Watkins JJ, McCarthy TJ (1996) Polym Mater Sci Eng 74: 402
126. Odian G (1991) Principles of polymerization. Wiley, New York
127. Biddulph R, Plesch P (1960) J Chem Soc 3913
128. Pernecker T, Kennedy JP (1994) Polym Bull (Berlin) 32: 537
129. Pernecker T, Kennedy JP (1994) Polym Bull (Berlin) 33: 13
130. Deak G, Pernecker T, Kennedy JP (1994) Polym Bull (Berlin) 33: 259
131. Kennedy JP, Pernecker T (1994) U.S. Pat 5,376,744
132. Pernecker T, Deak G, Kennedy JP (1995) Polym Prepr (Am Chem Soc, Div Polym Chem) 36: 243
133. Yokota H, Kondo M, Kagiya T, Fukui K (1968) J Polym Sci: Part A-1 6: 425
134. Yokota H, Kondo M, Kagiya T, Fukui K (1968) J Polym Sci: Part A-1 6: 435
135. Kagiya T, Kondo M, Narita K, Fukui K (1969) Bull Chem Soc Jap 42: 1688
136. Yokota H, Kondo M, Kagiya T, Fukui K (1969) Bull Chem Soc Jap 42: 1412
137. Fukui K, Kagiya T, Yokota H, Toriuchi Y, Matsumi S (1969) U.S. Pat 3,446,776
138. Yokota H, Kondo M, Kagiya T, Fukui K (1967) J Polym Sci: Part A-1 5: 3129
139. Clark MR, DeSimone JM (1994) Polym Prepr (Am Chem Soc, Div Polym Chem) 35: 482
140. Higashimura T, Aoshima S (1989) Macromolecules 22: 1009
141. Rokicki A, Kuran W (1981) J Macromol Sci, Rev Macromol Chem C21: 135
142. Sumitomo Chemical Co. L, Seitetso Kagaku Co. L (1968) British Pat 1,135,671
143. Lebrun J-J, Sagi F, Soria M (1986) Eur Pat Appl 0221824A1
144. Mistele CD, Thorp HH, DeSimone JM (1995) Polym Prepr (Am Chem Soc, Div Polym Chem) 36: 507
145. Mistele CD, Thorp HH, DeSimone JM (1996) J Macromol Sci A33: 953
146. Burke ALC, Maier G, DeSimone JM (1996) Polym Mater Sci Eng 74: 248
147. Romack TJ, Kipp BE, DeSimone JM (1995) Macromolecules 28: 8432
148. Kipp BE, Romack TJ, DeSimone JM (1996) Polym Mater Sci Eng 74: 264

Editor: Prof. Salamone
Received April 1996

Design and Construction of Supramolecular Architectures Consisting of Cyclodextrins and Polymers

Akira Harada

Department of Macromolecular Science, Faculty of Science, Osaka University, Toyonaka, Osaka, 560 Japan. *E-mail: harada@chem.sci.osaka-u.ac.jp*

Cyclodextrins (CDs) interact with various polymers of high specificity to give crystalline inclusion complexes in which CD molecules are threaded by a polymer chain. For example, α-CD forms such complexes with poly(ethylene glycol) (PEG) when the PEG molecular weight is higher than 200, but β-CD does not. On the other hand, for poly(propylene glycol) (PPG), the reverse is the case. Although both α and β-CD fail, γ-CD complexes with poly(methyl vinyl ether) (PMVE). In addition, α and γ-CD complex with oligo(ethylene) (OE) and poly(isobutylene) (PIB), respectively. Three quarters of the present paper describe the synthesis, structure, and properties of these CD-polymer inclusion complexes. The remaining part is concerned with rotaxanes including polyrotaxanes and also catenanes. A particularly detailed description is made concerning the work of the author and his coworkers which led to the synthesis of polyrotaxanes consisting of α-CD and PEG and the making of molecular tubes from them. It is expected that such tubes may find many useful applications in chemistry and biology.

Advances in Polymer Science, Vol. 133
© Springer-Verlag Berlin Heidelberg 1997

1 Introduction

In recent years, much attention has been focused on molecular recognition of low molecular weight compounds [1]. Molecular recognition is defined by the energy and the information involved in the binding and selection of substrates by a given receptor molecule. Typical hosts are crown ethers [2], cryptands [3], cyclophanes [4], and calixarenes [5]. However, the guests recognized by these hosts have been limited to small molecules and simple ions. Therefore, it has been inviting to find host molecules which can recognize and respond sensitively to larger and more complicated compounds including polymers. In biological systems, such as enzyme-substrate complexes, antigen-antibody complexes, DNA, RNA, and cell adhesion systems, macromolecular recognition, which means the recognition of a macromolecule by a macromolecule, plays a role in constructing supramolecular structures, achieving their specific functions, and maintaining their lives [6]. Yet, no approach to it with artificial host-guest systems has been foreseen.

Recently, it has become clear to the author that cyclodextrin is one of the promising hosts for macromolecular recognition, with the finding that the cylindrical channel formed when it is stacked in a linear row is able to accommodate one or more long-chain guests. Thus, he and his coworkers launched a series of experiments to explore the formation of inclusion complexes between

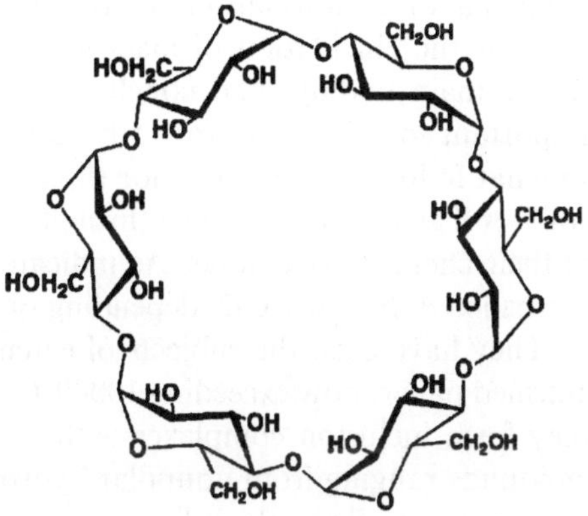

Figure 1.

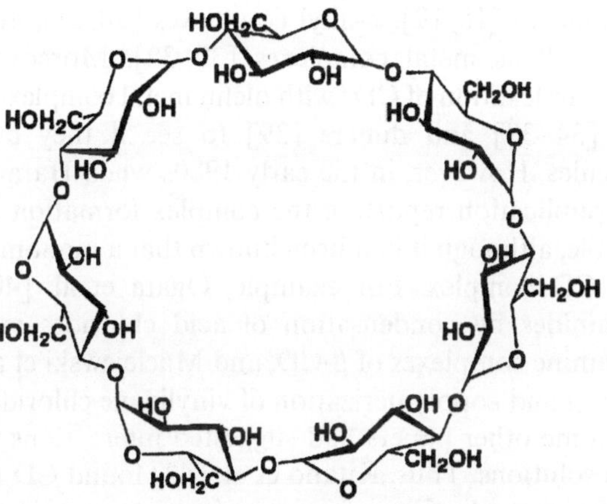

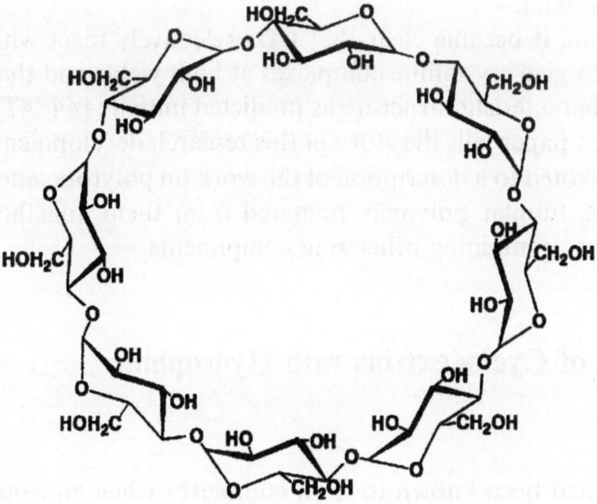

CDs and polymers, and have accumulated a lot of information that they believe is basically significant and interesting for the elucidation of macromolecular recognition. This paper summarizes the main findings and conclusions from their efforts, along with the related important contributions from other authors.

Cyclodextrin (abbreviated CD in what follows) is the name for three cyclic oligosaccharides, each containing 6 or 7 or 8 glucose units linked through α-1,4-glycosidic linkages. Fig. 1 illustrates their chemical structures. As indicated in the figure, these three compounds are called α, β, and γ-CD, depending on the numbers 6, 7 and 8 of glucose units. They have been the subjects of extensive investigation, with the number of published papers now exceeding 10000. One of the most notable features is that they form inclusion complexes with a wide variety of low molecular weight compounds ranging from nonpolar hydrocarbons to polar carboxylic acids and amines [7–9]. Recently, it has become clear that they also give inclusion complexes with organometallic compounds as ferrocenes [10–17], arene complexes [18, 19], π-allyl complexes [20, 21], and olefin complexes [22–29] as well as metal complexes [30–32]. Moreover, Stoddart et al. [33] found the complexation of CDs with olefin metal complexes. We prepared CD polymers [34–38] and dimers [39] to see if they can cooperatively bind long molecules. However, in the early 1980s when Harada et al. began their studies, no publication reporting the complex formation of CDs with polymers was available, although it had been known that a monomer polymerizes in situ within a CD complex. For example, Ogata et al. [40] succeeded in obtaining polyamides by condensation of acid chlorides and inclusion of hexamethylene diamine complexes of β-CD, and Maciejewski et al. [41] reported the polymerization and copolymerization of vinylidene chlorides forming adducts with β-CD. Some other papers had suggested interactions of CD with polymers in aqueous solutions. Thus, Kitano et al. [42] found CD to have effect on the critical micelle concentrations of some surfactants, and Iijima et al. [43] noted CD-polymer interactions in a study of diffusion of CD in water containing poly(styrenesulfonate).

As Harada et al. went on, it became clear that CDs selectively react with various kinds of polymers to give crystalline complexes at high yields and that those complexes have the characteristic structure as predicted initially [44–47]. Three quarters of the present paper tells the story of this research development, and the remaining part is devoted to a description of the work on polyrotaxanes entrapping many CD rings, tubular polymers prepared from them, together with rotaxanes and catenanes containing other ring components.

2 Complex Formation of Cyclodextrins with Hydrophilic Polymers [51–58]

As mentioned above, CDs had been known to form complexes when aqueous solutions of water-soluble low molecular weight compounds were added to their saturated aqueous solutions at room temperature. Harada and coworkers started examining whether the same happens between CDs and water-soluble nonionic polymers. First of all they found that CDs failed to complex with poly(vinyl alcohol) (PVA) and polyacrylamide(PAAm), but α-CD formed a complex with poly(ethylene glycol) (PEG) at as high a yield as shown in Table 1. Note that PEG has a smaller cross-section than PVA and PAAm. This table also shows that poly(propylene glycol) (PPG), which has a larger cross-section than PEG, can complex with β-CD and γ-CD at comparable high yields, but cannot do so with α-CD. The geometry worth noting here is that the cavity diameters of β and γ-CD, which are 7.0 and 8.5 Å, respectively, are bigger than 4.7 Å for the α-CD cavity. Another indication of Table 1 is that poly(methyl vinyl ether), which is the same in composition as PPG but has a methoxy group as the side chain, can complex with γ-CD, though not with the other two CDs. Thus, it was apparent that the size of the CD's cavity relative to the polymer's cross-sectional area plays an important role in their complex formation.

2.1 Complex Formation of CD with PEG [45, 47]

Addition of an aqueous solution of PEG to a saturated aqueous solution of α-CD at room temperature did not lead to complex formation unless the average molecular weight of PEG exceeded 200 [46]. Moreover, carbohydrate polymers such as dextran and pullulan failed to precipitate complexes with PEG, and the same was true for amylose, glucose, methyl glucose, maltose, maltotriose, cyclodextrin derivatives, such as glucosyl-α-CD and maltosyl-α-CD, and water-soluble polymers of α-CD crosslinked by epichlorohydrin. These facts suggested to Harada et al. the direction for further research.

Table 1. Comparison of hydrophilic polymers with various chain cross-sectional areas in formation of crystalline complexes with cyclodextrins

Polymer	Structure	Yield (%)		
		α-CD[a]	β-CD[b]	γ-CD[c]
PEG (MW 1000)	$+CH_2CH_2O+_n$	92	0	trace
PPG (MW 1000)	$+CH_2CHO+_n$ $\quad\quad\quad CH_3$	0	96	80
PMVE (MW 20000)	$+CH_2CH+_n$ $\quad\quad\quad O$ $\quad\quad\quad CH_3$	0	0	67

[a] α-CD saturated aqueous solution, 1.5 mL; polymers, 15 mg.
[b] β-CD saturated aqueous solution, 7 mL; polymers, 15 mg.
[c] γ-CD saturated aqueous solution, 2 mL; polymers, 15 mg.

2.1.1 Rates of Complex Formation

Because of the above finding Harada et al. noted it important to examine the rate of complex formation between α-CD and PEG as a function of PEG molecular weight, and results obtained are illustrated in Fig. 2, where the rate of turbidity development after mixing fixed amounts of α-CD and PEG is plotted against PEG molecular weight. It is seen that the rate attains a maximum at a PEG molecular weight near 1000. The maximum may be attributed to the competition of two effects: the increased chance for the CD to find polymer chains in its neighbor and the decrease in the number of chain ends with the increase in chain length. Aqueous mixtures of PEG and β-CD exhibited no turbidity change at any PEG molecular weights.

2.1.2 Molecular Weight Effects on the Yield of Complexes

With the α-CD-PEG complexes isolated by filtration or centrifugation, the yields of complexes were determined as a function of PEG molecular weight, assuming that two ethylene glycol units are bound to one α-CD, i.e., the stoichiometric ratio is 2 : 1, and the results shown in Fig. 3 were obtained. It is

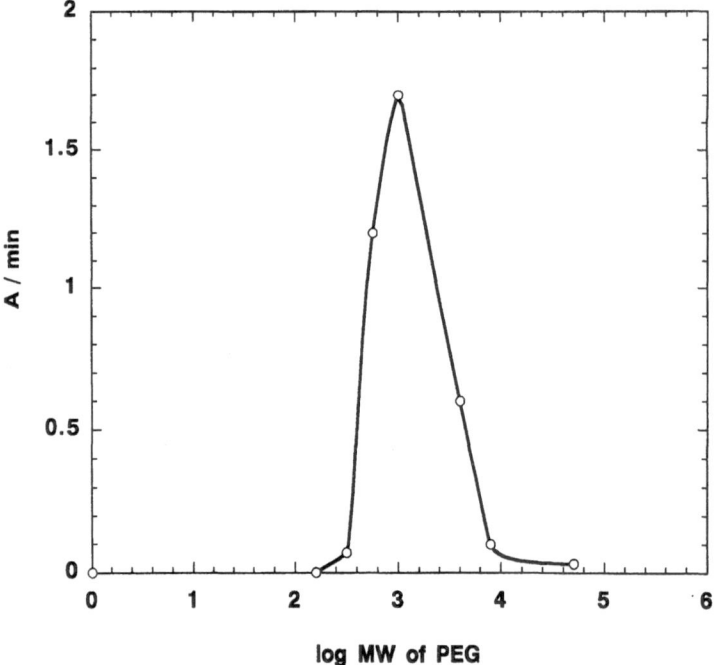

Figure 2.

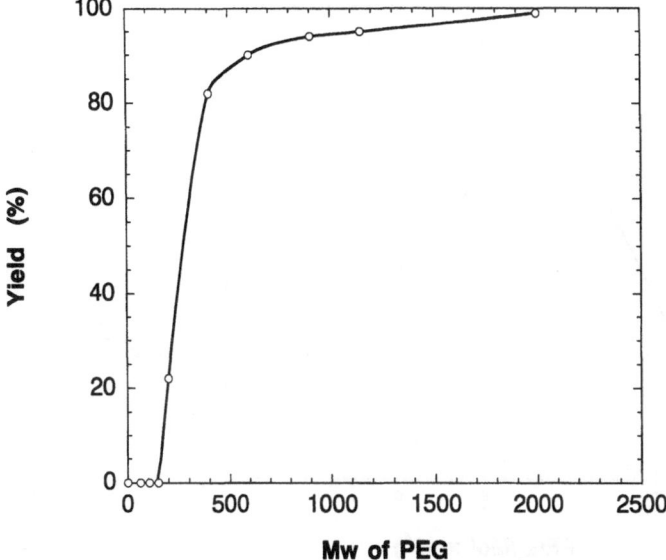

Figure 3.

seen that no complex formation occurs until the PEG molecular weight reaches about 200, but once complexation starts, the yield sharply increases with increasing the molecular weight, and becomes almost quantitative at the molecular weights above 1000. The complex formation was initially slow when the molecular weight was higher than 1000, but equilibrium was attained after the solution had been left standing for several hours.

The above finding led to another interesting conclusion that a minimum chain length is essential for PEG to form a stable complex with α-CD. The chain length selectivity implies that some cooperative action is involved in this complex formation as is well known in the complex formation of PEG with hydrogen-donor polymers such as poly(acrylic acid).

2.1.3 Stoichiometries of Complex Formation

Next, with a PEG sample having as an average molecular weight of 600, the yields of complexes with α-CD were measured as a function of the amount of PEG in the solution, and the results presented in Fig. 4 were obtained. It is seen that the yield initially increases linearly with the polymer amount and begins to level off at a molar ratio of about 2:1 (ethylene glycol:α-CD). The saturation corresponds to the consumption of α-CD of more than 90%. Fig. 5 delineates the data obtained by varying the amount of α-CD and keeping that of PEG fixed. It also proves that the complex formation obeys the 2:1 stoichiometric ratio. Further support to this fact can be seen from the ^{1}H NMR spectra shown in

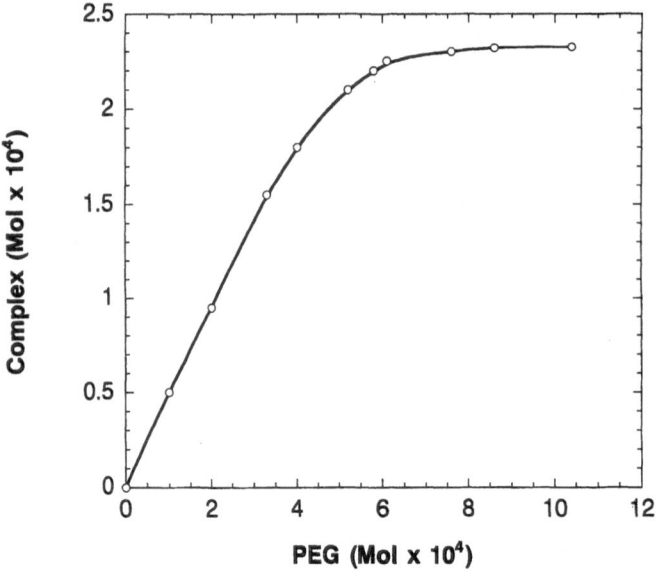

Figure 4.

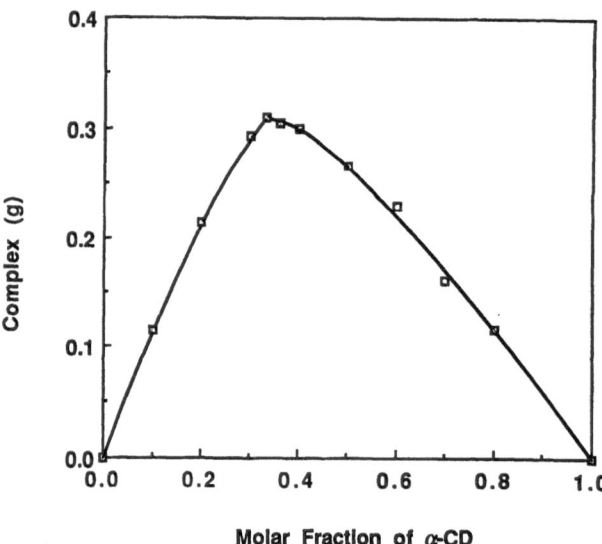

Figure 5.

Fig. 6. Thus, the conclusion is that the 2:1 stoichiometry governs the complex formation of α-CD with PEG, regardless of the mixing ratio of the two components. In this connection, the geometry, namely that the length of two ethylene glycol units (6.6 Å) is essentially equal to the depth of the α-CD cavity (6.7 Å), is worth mentioning.

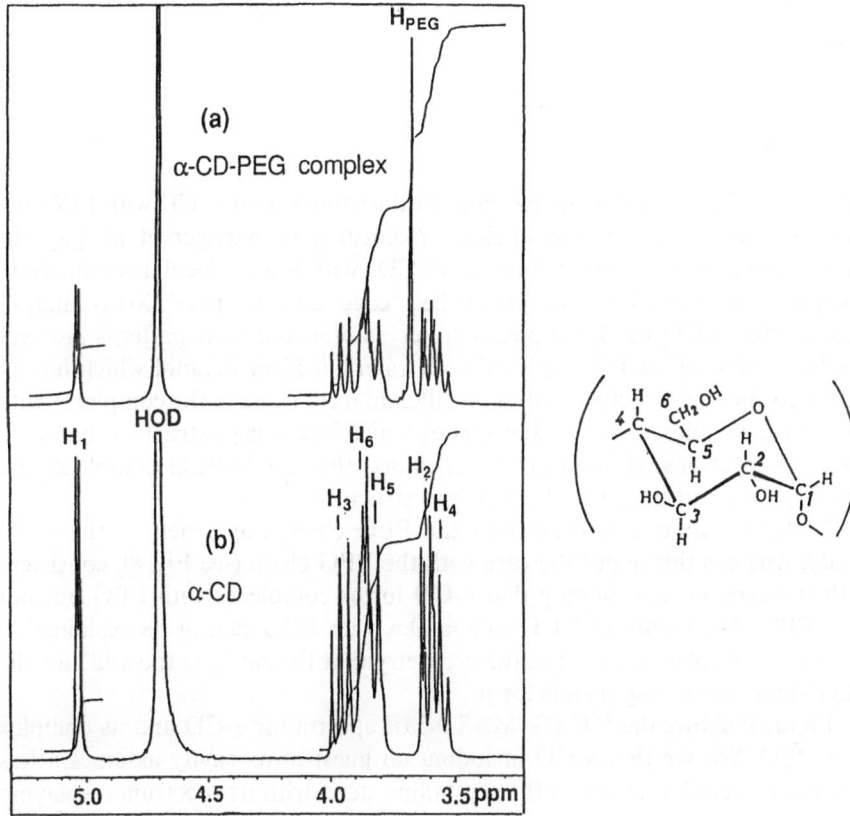

Figure 6.

2.1.4 Properties of the Complexes

The α-CD-PEG complexes are crystalline solids if the PEG molecular weight is lower than 2000. They are water-soluble for the molecular weights below 1000, but heating is needed to make those of higher molecular weight dissolve in water. Addition of an excess of a low molecular weight compound, such as benzoic acid, propionic acid, and propanol, stabilizes the suspension if the PEG molecular weight is as low as 1000. The complex formation occurs reversibly, and the equilibrium is not affected by the addition of salts such as NaCl and KCl. Thus no ionic interaction appears to be involved in the complex formation between α-CD and PEG. However, since the addition of urea, a known hydrogen-bond breaker, solubilizes the complexes, it is certain that hydrogen bonding plays a role in this complex formation.

The decomposition temperatures of the complexes were a little higher than that for a α-CD itself. For example, when PEG had a molecular weight of 1000 the decomposition occurred above 300 °C, while pure α-CD melted and decomposed below it. This fact shows that α-CD is stabilized by complexation with PEG.

2.1.5 Inclusion Modes

Figure 7 shows X-ray powder patterns for the complexes of α-CD with PEG and some low molecular weight analogs. According to Saenger et al. [9], the structures of the inclusion complexes of CDs with low molecular weight compounds are classified into two types called "cage" and "channel". X-ray analysis showed that α-CD-PEG complexes are crystalline and their patterns are very similar to that of α-CD complexed with valeric acid or octanol which has an extended channel structure, but quite different from those of the complexes with acetic acid, propionic acid, and propanol, which have a cage structure. Based on these observations, Harada et al. concluded that α-CD-PEG complexes are isomorphous and have the channel-type structure.

Molecular models indicate that the PEG chain can penetrate the α-CD cavity, whereas this is not the case with the PPG chain (see Fig. 8), consistent with the experimental finding that α-CD forms complexes with PEG but not with PPG. The failure of β-CD to complex with PEG cannot be explained in terms of molecular models, but the geometry that the cavity is too wide for the PEG chain seems responsible for it.

Figure 9 shows the ^{13}C CP/MAS NMR spectra for α-CD and its complex with PEG. We see that α-CD including no guest in its cavity assumes a less symmetrical conformation in the crystalline state, with its spectrum displaying resolved C-1 and C-4 resonances associated with the six α-1,4-linked glucose residues. Note that the peaks at 80 and 89 ppm, corresponding respectively to the C-1 and C-4 adjacent to the conformally strained glucose linkage, disappear in the spectrum of the complex, in which each carbon of glucose appears as a single peak. Thus, it appears that, in the complex with PEG, α-CD assumes a symmetrical conformation and all of its glucose units find themselves in a similar environment.

2.1.6 Complex Formation of α-CD with Monodisperse Oligo(ethylene glycol)s [59]

The initial experiments of Harada and coworkers were done using commercially available PEG, so that the complexes obtained were undoubtedly polydisperse and heterogeneous. In order to see if the early finding that α-CD made no complex with such low molecular members of PEG as ethylene glycol and bis(ethylene glycol) was due to a sample's polydispersity, they went one step further by preparing a series of monodisperse oligo(ethylene glycols) (OEG), i.e., $HO(-CH_2CH_2O)_nOH$ (for n = 8, 12, 18, 20, 28, 36, 44), using the Bomer method [60] followed by repeated purification by preparative size-exclusion

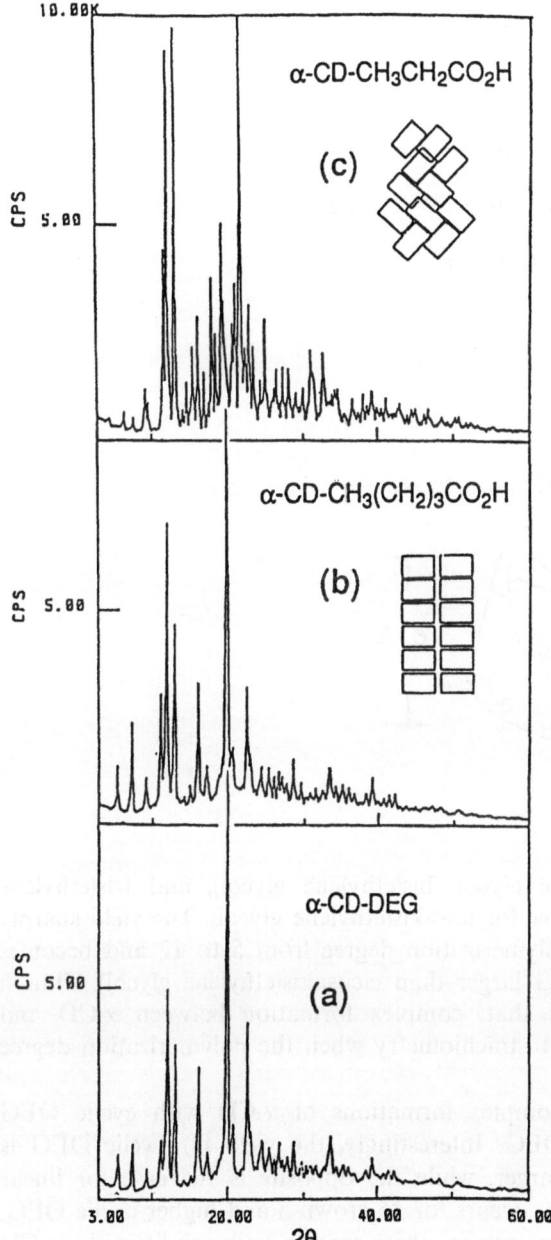

Figure 7.

chromatography. The yields of complexes of α-CD with these OEG appear plotted against the polymerization degree of OEG in Fig. 10; their values were calculated assuming the 2:1 stoichiometry that has been established in a previous section. This graph convinces us that complex formation with α-CD does

PEG(Polyethylene glycol)

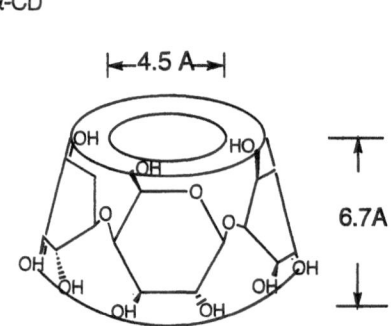

α-CD

Figure 8.

not take place for ethylene glycol, bis(ethylene glycol), and tris(ethylene glycol) but becomes detectable for tetrakis(ethylene glycol). The yield sharply increases in the range of polymerization degree from 5 to 12 and becomes almost quantitative for OEG larger than eicosakis(ethylene glycol). Thus a more definite conclusion is that: complex formation between α-CD and PEG begins to obey the 2:1 stoichiometry when the polymerization degree exceeds 6.

Table 2 compares the complex formations of α-CD with cyclic OEG (crown ethers) and linear OEG. Interestingly, the yield for cyclic OEG is lowered as the guest gets larger, while the opposite is the case for linear OEG. No complex formation occurs for 15-crown-5 and higher cyclic OEG. This is probably because these crown ethers are too bulky to fit in the α-CD cavity. Evidently, having no chain ends, the α-CD molecules cannot intercross with cyclic OEG chains. Dioxane gave a one-to-one complex and 12-crown-4 formed a two-to-one complex with α-CD. These facts imply that effective end groups are needed for allowing linear OEG chains to be recognized by α-CD. So Harada and coworkers proceeded to investigating PEG modified at its chain ends.

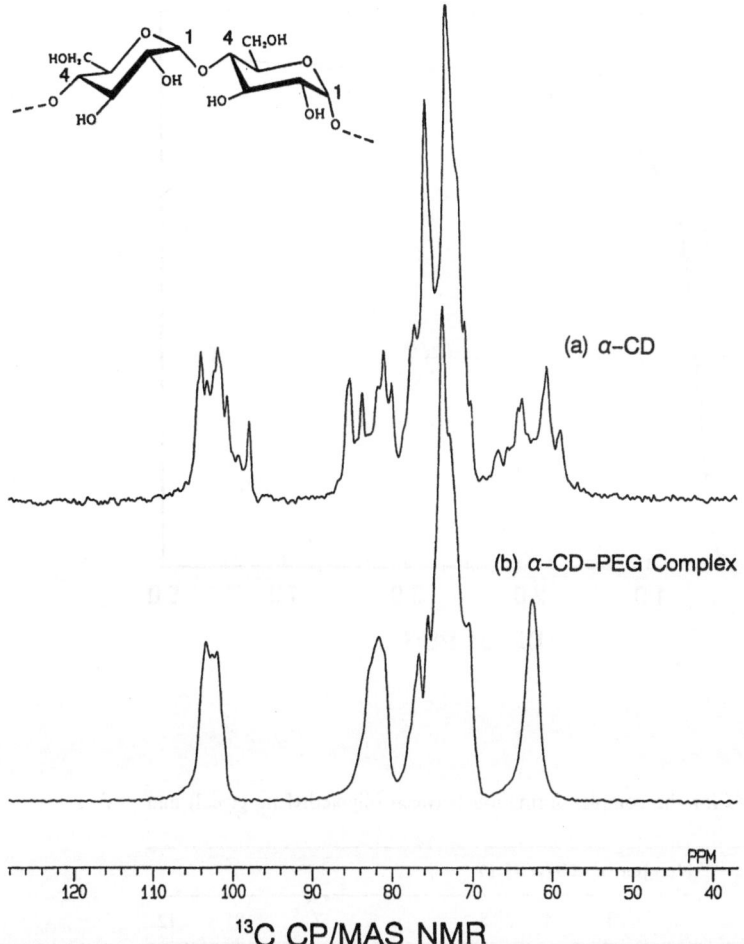

Figure 9.

2.1.7 Complex Formation between α-CD and PEG Derivatives

Table 3 shows the experimental results obtained with the PEG. First, PEG carrying small end groups, such as methyl, dimethyl, and amino, can give complexes at higher yields than for unmodified PEG. This implies that hydrogen bonding between the OH groups of PEG and α-CD plays no essential role in the complex formation. Next, PEG carrying bulky end groups, such as 3,5-dinitrobenzoyl and 2,4-dinitrophenyl, fail to complex with α-CD, probably because these substituents are too large to allow the chain to enter the α-CD cavity.

Figure 11 proposes a structure for the PEG-α-CD complex on the basis of the conclusions derived so far. In that, a single PEG chain penetrates through

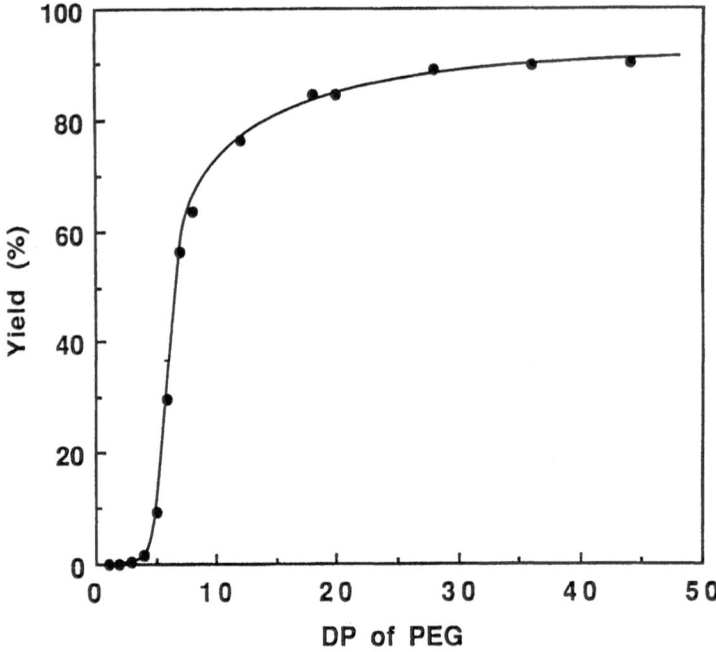

Figure 10.

Table 2. Yields for the complex formation between oligo(ethylene glycol) and α-CD

	Number of –CH₂CH₂O–							
	2	3	4	5	6	7	8	12
Linear (%)	0	0	2	9	30	56	64	76
Cyclic (%)	21	–	9	0	0	–	–	–

the channel formed by the cavities of α-CD molecules bound together by hydrogen bonds in the head-to-head and tail-to-tail fashion, with the secondary hydroxy group side as the "head" and the primary hydroxy group side as the "tail". Including PEG in the α-CD channel is entropically not favourable, but hydrogen bond formation between adjacent cyclodextrins promotes it since a lowering of enthalpy results. Thus, the structure as in Fig. 11 should be highly probable when the latter effect is overwhelming.

Table 3. Complex formation between CD and PEG with various end groups

R(CH₂CH₂O)ₙCH₂CH₂R'			Yield (%)	
R	R'	MW	α-CD	β-CD
–OH	–OH	1000	90	0
–NH₂	–NH₂	1450	90	0
–OCH₃	–OCH₃	1000	93	0
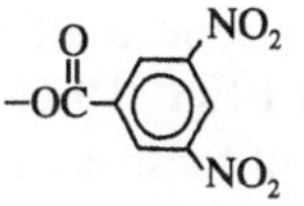 $-O\overset{O}{\underset{\parallel}{C}}\text{-}\langle\text{NO}_2\rangle_2$	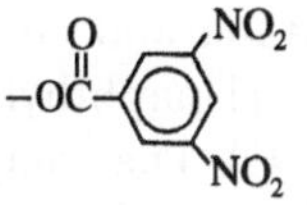 $-O\overset{O}{\underset{\parallel}{C}}\text{-}\langle\text{NO}_2\rangle_2$	900	0	0
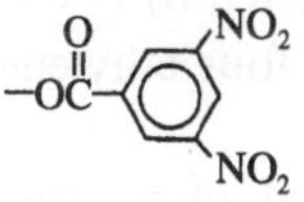 $-O\overset{O}{\underset{\parallel}{C}}\text{-}\langle\text{NO}_2\rangle_2$	–OCH₃	900	77	10
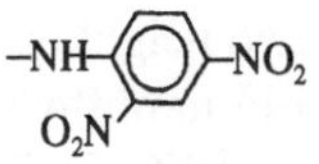 $-NH\text{-}\langle\rangle\text{-}NO_2, O_2N$	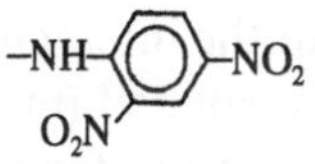 $-NH\text{-}\langle\rangle\text{-}NO_2, O_2N$	3700	0	0

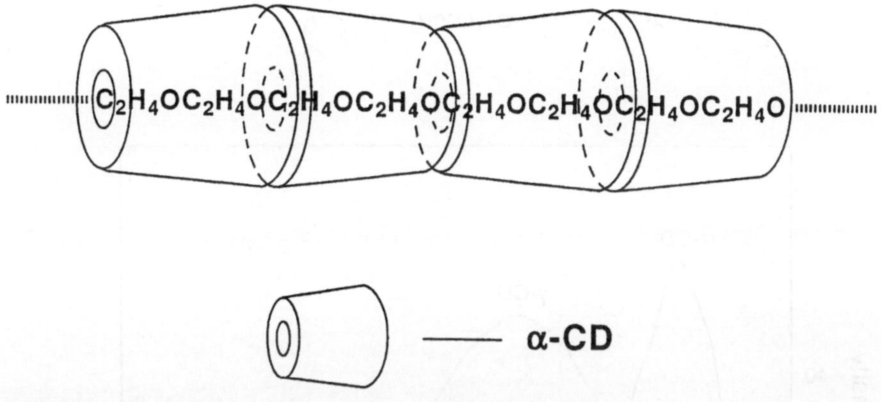

Figure 11.

2.1.8 Double-chain Inclusion Complexes of γ-CD with PEG [61]

Though γ-CD was found to form complexes also with PEG, the yields were too small for their characterization. Thus, Harada et al. turned to two PEG derivatives-bis(3,5-dinitrobenzoyl)-PEG (PEG-DNB2) and bis(2,4-dinitrophenyl-amino)-PEG (PEG-DNP2)-, which were found to complex with this CD at high yields. Using a fluorescent probe technique the resulting complexes were shown to include four ethylene glycol units in one γ-CD cavity. However, these PEG derivatives formed no complex with α-CD. This is attributable to the fact that their chains are too thick to enter the slim α-CD cavity.

To obtain additional information PEG-DNB2 was allowed to complex with
γ-CD first by adding it to an aqueous solution of γ-CD at the 2 : 1 stoichiometric
ratio and then further adding the same amount of it to the mixture; in total, 90%
of the CD was consumed for complexation. The area of the ^{1}H NMR peak for
CD relative to that for PEG in the isolated complex was consistent with the
above finding. It is interesting to compare this result with the complexation of
α-CD with PEG, by the fluorescent probe technique.

In order to work out in more detail Harada and coworkers took a step
further by preparing fluorescence-labeled bis(1-naphthylacetyl)-PEG (PEG-
1N2) and bis(2-naphthylacetyl)-PEG (PEG-2N2). The complexes of these compounds with γ-CD were isolated by standard methods and subjected to NMR
measurement. The results also conformed to the conclusion that four ethylene
glycol units are included in one γ-CD.

The emission spectrum of the PEG-2N2 complex consisted of a large
contribution from a collective excitation due to the interaction of two neighboring naphthyls and a small contribution from isolated (monomeric) naphthyls
(see Fig. 12). This differs from the emission spectrum for the α-CD-PEG-2N2,

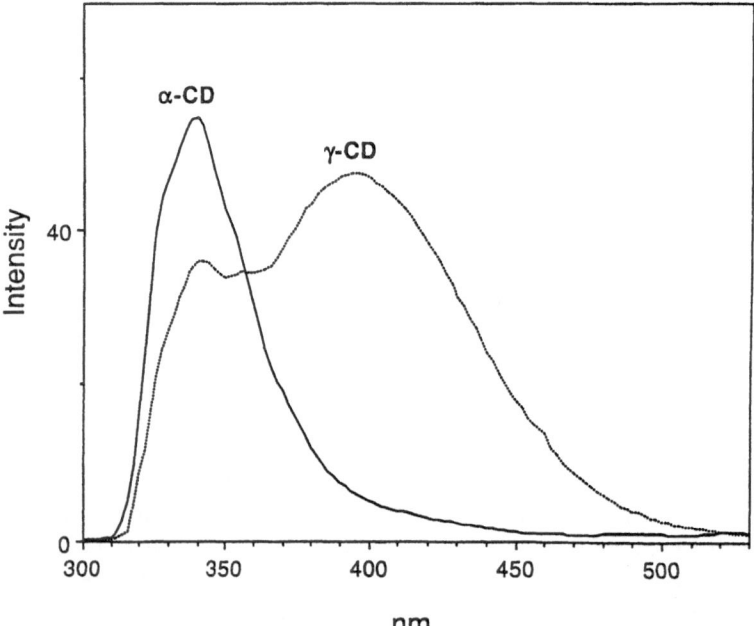

Figure 12.

which exhibits only the emission from isolated naphthyls. These results may also be taken to prove that the α-CD cavity is threaded by a single PEG chain and the γ-CD cavity by two. Spectral data on the complexes of PEG-1N2 with CDs confirmed this conclusion. In the absence of CD, only a small monomer emission contributed by excimers was observed. Addition of β or α-CD caused the monomer emission to increase and the excimer emission to decrease. When γ-CD was present, the emission was primarily from excimers, which indicates that two naphthyl groups were involved in the γ-CD cavity.

We may envisage the two modes of chain inclusion as sketched in Fig. 13: in one, a linear stack of CDs is threaded by two separate chains, and in the other it is threaded by one folded chain. To see which is more likely to occur, Harada and coworkers prepared monosubstituted naphthyl derivatives of PEG (PEG-2N) and found that the emission spectra for their complexes with γ-CD were

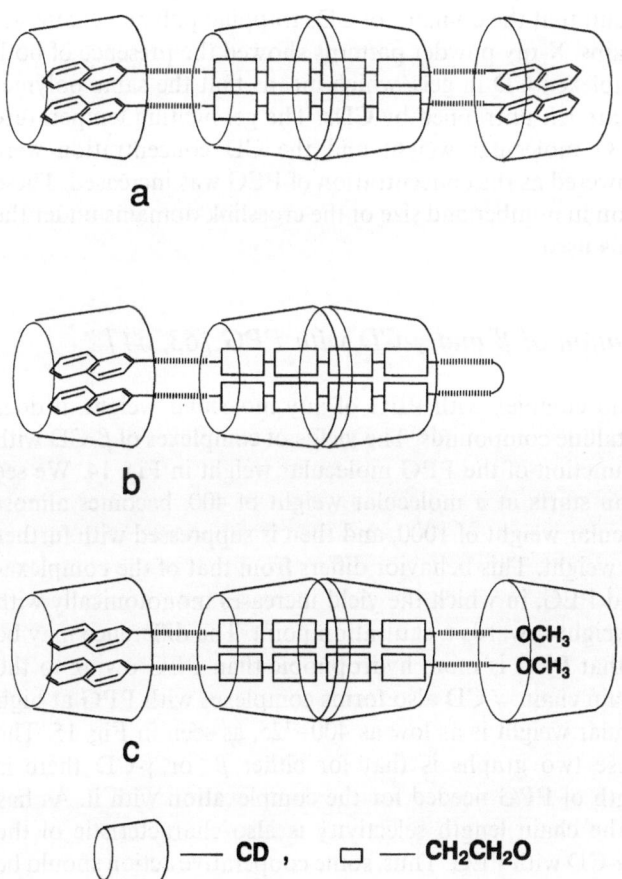

Figure 13.

similar to those for the disubstituted forms. This convinced them that the two separate chain inclusion is more probable.

In passing, we again touch upon some geometrical facts. The diameter of the γ-CD is 8.5–9 Å, which is about twice as large as that of the α-CD. The cavity depth is 7 Å for all α, β and γ-CD, and almost equals the length of two ethylene glycol units. Molecular models show that the γ-CD cavity is wide enough to accommodate two PEG chains, but the α-CD cavity is too narrow to make this possible.

2.1.9 Gel Formation between α-CD and High Molecular Weight PEG [62]

Gels covering a wide range of concentration were formed when high molecular weight PEG and α-CD were mixed at different ratios in water. The gelation was enhanced by increasing the concentrations of the two components, and also by an increase in the molecular weight of the PEG. These facts suggest that gelation takes place with the result that the domains of CDs trapping polymer chains act as crosslinks of the chains. X-ray powder patterns showed the presence of both complexed and uncomplexed CD in gels, which means that the same polymer chains in the solution are left untrapped by CDs. The gel-melting temperature got higher as the PEG molecular weight and the CD concentration were increased, but it was lowered as the concentration of PEG was increased. These facts reflect the variation in number and size of the crosslink domains under the experimental conditions used.

2.2 Complex Formation of β and γ-CD with PPG [63, 64]

Though β-CD forms no complex with PEG of any molecular weight, it does with PPG to give crystalline compounds. The yields of complexes of β-CD with PPG are shown as a function of the PPG molecular weight in Fig. 14. We see that complex formation starts at a molecular weight of 400, becomes almost quantitative at a molecular weight of 1000, and then is suppressed with further increases in molecular weight. This behavior differs from that of the complexation between α-CD and PEG, in which the yield increases monotonically with increasing molecular weight towards a saturation point. The difference may be attributed to the fact that PPG is more hydrophobic than PEG owing to the methyl group on its main chain. γ-CD also forms complexes with PPG at high yields when the molecular weight is as low as 400–725, as seen in Fig. 15. The point revealed by these two graphs is that for either β- or γ-CD there is a minimum chain length of PPG needed for the complexation with it. As has already been shown, the chain length selectivity is also characteristic of the complex formation of α-CD with PEG. Thus, some cooperative action should be involved in the complexation of CDs with hydrophilic polymers. For example, it is likely that the hydrogen bonding of CD rings is aided by neighboring polymer

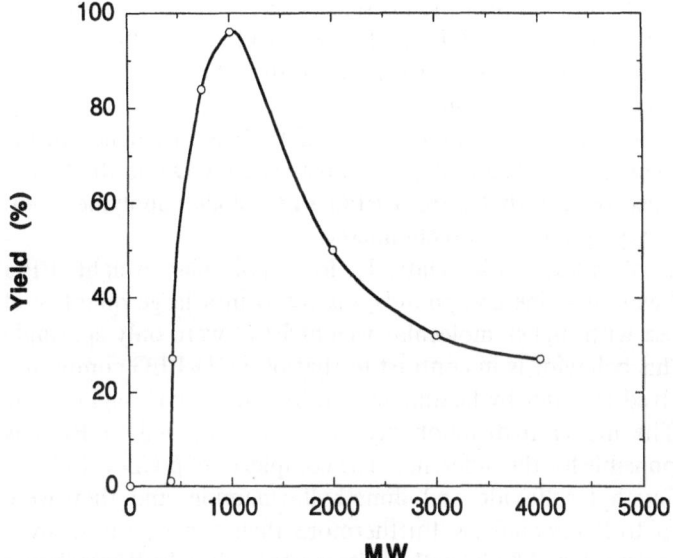

Figure 14.

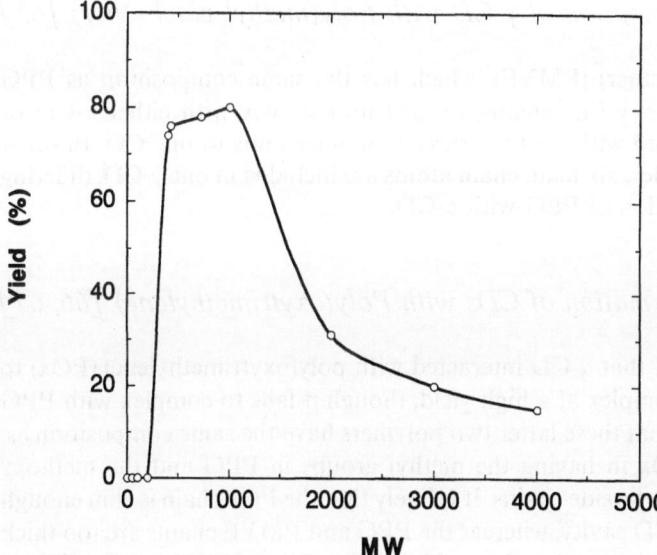

Figure 15.

chains, which carry many sites interacting with them. This view is consistent with the fact that PPG failed to complex with 2,6-di-O-methyl-β-CD, 2,3,6-tri-O-methyl-β-CD, and a water-soluble polymer of β-CD, all of which carry no hydroxyl group and hence cannot form hydrogen bonding with PPG.

Molecular model studies confirm the possibility of the PPG chain to penetrate the β-CD cavity, but not the α-CD cavity owing to the steric hindrance of the methyl groups on the chain. They also indicate that the β-CD cavity can accommodate two propylene glycol units.

^{1}H NMR data showed that the methyl peak of PPG is broadened in the presence of β-CD, which suggests an interaction between β-CD and the PPG's methyl group. We also remark that a PEG-PPG-PEG block copolymer complexed with β-CD obeying the 2:1 stoichiometry.

The complexes of β-CD with relatively low molecular weight PPG (400–700), isolated as crystalline compounds, dissolved in a large quantity of water, whereas those with higher molecular weight PPG were only sparingly soluble in water. This behavior is in contrast to that of α-CD-PEG complexes, which can be dissolved in water by heating even if the molecular weight of the polymer is high. The higher hydrophobicity of PPG compared to PEG is supposed to be responsible for the difference. The complexes of β-CD with PPG were soluble in dimethyl sulfoxide and dimethylformamide, and they were crystalline according to X-ray analysis. Furthermore, their X-ray patterns were similar to that of the complex of β-CD with p-nitroacetanilide, which was shown to have a channel-type structure from a single crystal X-ray study.

2.3 Complex Formation of γ-CD with Poly(methyl vinyl ether) [65]

Poly(methyl vinyl ether) (PMVE), which has the same composition as PPG except for its methoxy side chains, formed no complex with either α-CD or β-CD, but complexed with γ-CD at three monomer units to one CD. In other words, in this complex, six main chain atoms are included in one γ-CD, differing from the complexation of PEG with α-CD.

2.4 Complex Formation of CDs with Poly(oxytrimethylene) [66, 67]

Harada et al. found that α-CD interacted with poly(oxytrimethylene) (POx) to give a crystalline complex at a high yield, though it fails to complex with PPG and PMVE. Note that these latter two polymers have the same composition as, but differ from, POx in having the methyl groups in PPG and the methoxy groups in PMVE as the side chains. It is likely that the POx chain is slim enough to penetrate the α-CD cavity, whereas the PPG and PMVE chains are too thick to do that. Fig. 16, delineating the yields of complexes of POx with α-CD as a function of polymer molecular weight, reveals a maximum yield to occur at about 1000 in molecular weight. The difference from the already-mentioned behavior of α-CD-PEG complexes is probably a reflection of a higher hydrophobicity of POx compared to PEG. β-CD was also found to form crystalline inclusion complexes with POx, but not with PEG. The reason for the difference would be not only the more hydrophobic nature of POx compared to PEG but

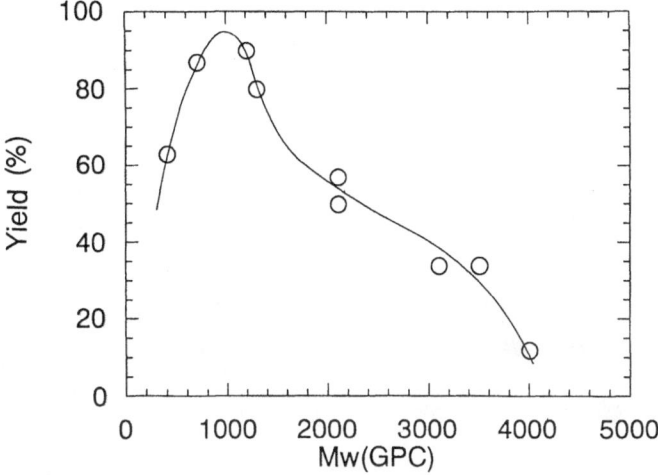

Figure 16.

also the fact that the POx chain can take conformations other than planar zigzag, which allow the chain to accommodate in a wider cavity of β-CD. Another notable feature of Fig. 16 is that the yields of complexation with POx are definitely larger for α-CD than for β-CD over the entire molecular weight range examined. This implies that the POx chain is more nicely included in the cavity of α-CD than in that of β-CD.

The inability of γ-CD to complex with POx under the same conditions as for β-CD indicates the failure of POx to meet the prerequisite for inclusion complex formation: that the polymer chain fits well into the CD cavity and is hydrophobic enough to stabilize the complex.

According to conformational energy calculations, the POx chain can assume conformations other than planar zigzag, such as T_3G and T_2G_2, and, according X-ray and IR studies, these two conformations are stable. The POx assuming them is thicker than the planar zigzag one, so that it may be included in β-CD as well as α-CD.

The complex formation of α-CD with POx was found to obey the same stoichiometry as that for PEG. In fact, the X-ray powder pattern of the complex does not differ from that of the α-CD-PEG complex. Fig. 17(a) illustrates the POx chain in the complex with α-CD when its conformation is planar zigzag, and Fig. 17(b) sketches the POx chain in the β-CD cavity where it favors the T_3G conformation thanks to the wider cavity of this CD.

2.5 Complex Formation of CDs with Poly(tetrahydrofuran) [68]

CDs were found to form inclusion complexes also with poly(tetrahydrofuran) (PTHF) of different molecular weights. In particular, γ-CD gave the complexes

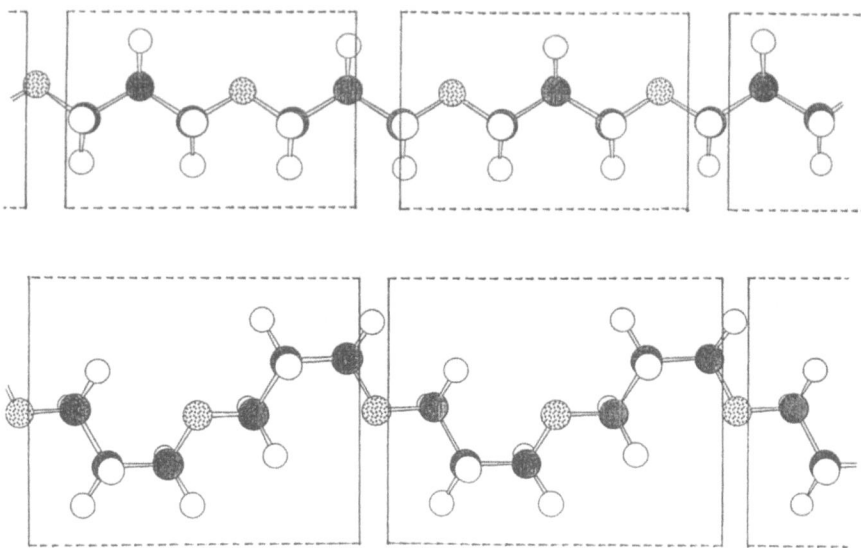

Figure 17.

at high yields. However, it failed to complex with poly(oxytrimethylene), despite the
fact that the monomer unit of this polymer differs from that of PTHF only by one
methylene group. The yields of complexes of PTHF with α-CD decreased with
increasing polymer molecular weight, but with γ-CD they first increased, reached
a maximum at a molecular weight around 1000, and then decreased with increasing
polymer molecular weight. The stoichiometric ratio was found to be 1:1.5, and the
length of PTHF included in a single α-CD cavity did not differ from that of PEG in
the same CD. Furthermore, X-ray and solid state NMR studies showed the
complexes of PTHF with α- and γ-CD to have the channel-type structure.

According to Steiner et al. [69], who studied the complex of β-CD with
1,4-butanediol, a monomer model of PTHF, the crystal packing in this complex
is the cage type and isomorphous to that of the β-CD hydrate, and the methylene
chain vibrates in the cavity. However, in the β-CD-PTHF complex, the polymer
chain lays fixed inside a column formed by linearly bonded CD molecules. In
agreement with this picture, CP/MAS NMR spectra showed the PTHF chain in
the complex to be much less flexible than that in its mixture with the CD.

3 Complex Formation between CDs and Hydrophobic Polymers [70]

Harada et al. found that hydrophobic polymers, such as oligoethylene and
polyisobutylene, also form complexes with CDs, as illustrated in Table 4. Here
we observe the following.

Table 4. Formation of solid-state complexes between cyclodextrins and hydrophobic polymers/oligomers with various chain sectional area

Polymer/oligomer	structure	molecular weight	Yield (%)		
			α-CD	β-CD	γ-CD
OE (20)	—CH$_2$CH$_2$—	563	63	0	0
squalane	—CH$_2$CHCH$_2$CH$_2$— 　　　| 　　　CH$_3$	423	0	62	24
PIB	CH$_3$ 　　| —CH$_2$C— 　　| 　　CH$_3$	~ 800	0	8	90

α-CD gives a crystalline complex at a high yield with oligoethylene (OE), which has the smallest cross-sectional area among the three compounds in the table. Neither β nor γ-CD complexes with OE under the same experimental conditions. However, the former complexes with squalane, which has one methyl side chain per monomer unit, whereas the latter does so with polyisobutylene (PIB), which has a larger cross section than squalane. These results point out the important role that the relative sizes of polymer cross section and the CD cavity play in the complex formation of hydrophobic polymers or oligomers with CDs as well as in the case of hydrophilic polymers.

3.1 Complex Formation Between OE and α-CD [71]

When an aqueous mixture of α-CD and OE was heated to above the melting temperature of OE and then sonicated, it turned turbid and precipitated a crystalline complex which was found to include OE in the CD cavity. The same occurred when dimethylformamide (DMF) replaced water. In Fig. 18, the yields of α-CD-OE complexes in these two liquids are shown as a function of n, the polymerization degree of OE, calculated assuming the 3:1 stoichiometry. The yields in water are independent of n and higher than those in DMF; the latter varies with n in such a way that the yield becomes detectable at values of n larger than 6, attains a maximum, and decreases with increasing n. The declining behavior is due to the lowering in solubility and the enhanced difficulty in diffusing into the α-CD cavity of higher molecular weight OE chains in DMF. Another point to note is that, in water, no minimum chain length appears for the complexation of OE with α-CD. This lack of chain length selectivity is in contrast to the complexation of hydrophilic PEG with the same CD. A considerable solubility difference between the complexes of OE and PEG with α-CD may be responsible for it. Thus, the OE complexes cannot be dissolved in water even with

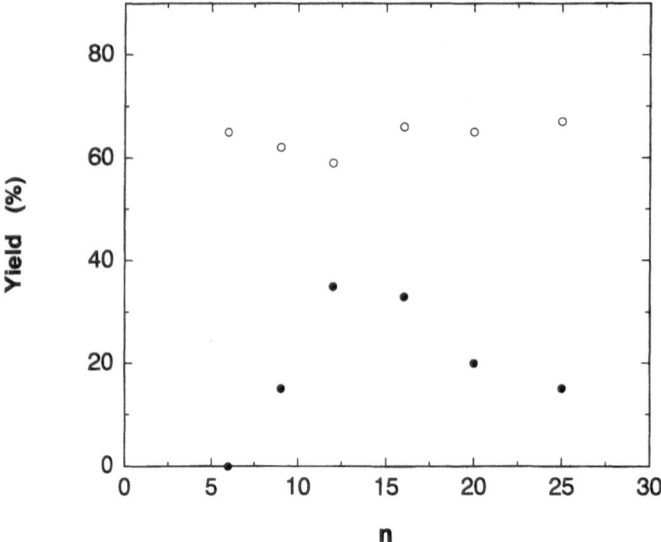

Figure 18.

heating and in most organic solvents except for DMF. In DMF, the solubility is still very low, but, upon heating, it increases to the degree that makes NMR spectra of the complex in DMF d_7 to be measured. The complex dissociated to its components in hot dimethylsulfoxide (DMSO).

In obtaining Fig. 18, it has been assumed that the complexation between α-CD and OE follows the 3 : 1 stoichiometry independently of the mixing ratio of the two components. In fact, ^{1}H NMR spectra of the isolated complexes confirmed this assumption, and furthermore, the complexation of α-CD with such derivatives of OE(6) (dodecane) as α,ω-diaminohexaethylene, α,ω-dihydroxyhexaethylene, and α,ω-dicarboxyhexaethylene was found to obey the 3 : 1 stoichiometry.

X-ray studies showed that all the complexes listed in Table 4 are crystalline and displayed the diffraction patterns as predicted from the molecular models, in which the α-CD cavity is threaded by an OE chain but not by a squalane chain, and also from the experimental finding that α-CD forms complexes with the former but not with the latter. The fact that neither β-CD nor γ-CD can complex with OE may be ascribed to the thickness of the OE chain that is too thin to interact effectively with the inner surfaces of these CD rings. The ^{13}C CP/MAS NMR spectra of the α-CD-OE complexes were similar to those of α-CD-PEG complexes, exhibiting each carbon atom of glucose as a single peak. Thus, the α-CD molecules in the complex with OE assume a symmetrical conformation and each glucose unit finds itself in a similar environment.

The solid state ^{13}C NMR signals of OE in its complexes with α-CD appeared separately from the spectra of α-CD-PEG complexes in a higher magnetic field, in which the signals of the two components were overlapped. This feature allow us to extract more information about the OE chain included in the CD cavity. Figure 19 compares the ^{13}C NMR spectra for a physical mixture of α-CD and OE(12) and their complex. Note that this solid-state NMR method gives stronger signals for the carbons in a given sample than ^{13}C CP/MAS. The spectrum for the physical mixture indicates that the peak intensity for the methyl ends of OE(12) is considerably stronger than that for its ethylene backbone. This can be taken as reflecting that, in the uncomplexed OE(12), the ethylene backbone is confined in a crystal lattice so that it is less mobile, while its methyl end groups have higher mobility. In Fig. 19, we see that each glucose carbon of α-CD appears as a single peak, as in the CP/NMR spectrum, and the peak intensity for the ethylene backbone is much stronger than that for its methyl ends. This implies that the ethylene backbone of OE(12) gains higher mobility when complexed with α-CD. The intensity ratios of the peaks for the OE backbone and α-CD are larger in the PST spectrum than in the CP one. This

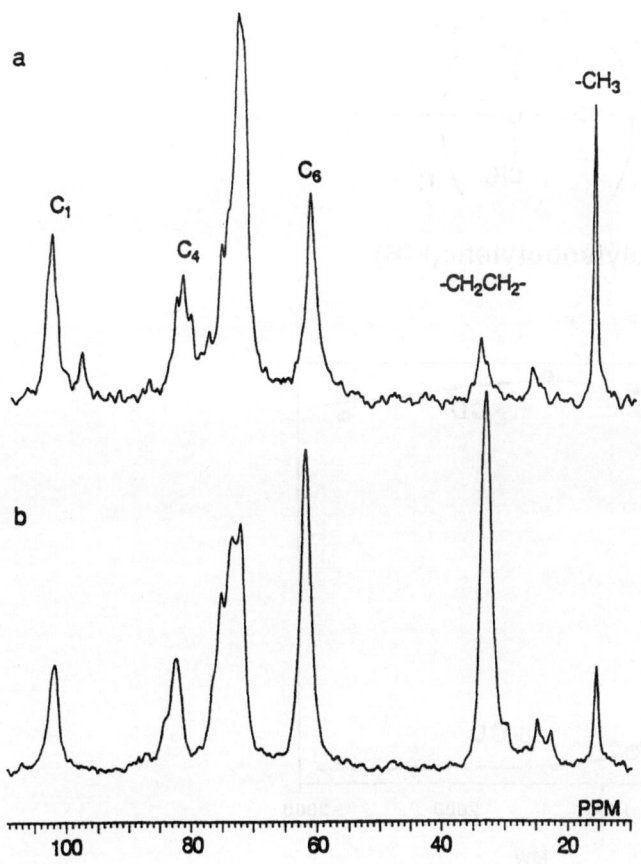

Figure 19.

suggests that, in the complex, the OE backbone is more mobile than α-CD. All these conclusions are in line with the picture that when complexed with α-CD, the OE chain is trapped in a tunnel formed by α-CD rings which arrange themselves in a crystal frame.

Yonemura et al. [72], Saito et al. [73], and Watanabe et al. [74] reported complex formation of α-CD with some charged derivatives of oligomethylene, but observed only the formation of one-to-one complexes in aqueous solutions, not of any crystalline one.

3.2 Complex Formation of Polyisobutylene with CDs [75]

When polyisobutylene (PIB) was added to an aqueous solution of either γ-CD or β-CD and then sonicated at room temperature, the mixture became turbid and precipitated a crystalline complex. Figure 20 shows how the complexation

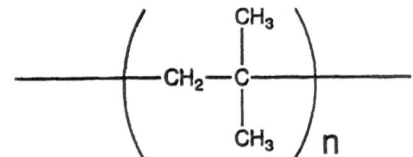

Polyisobutylene(PIB)

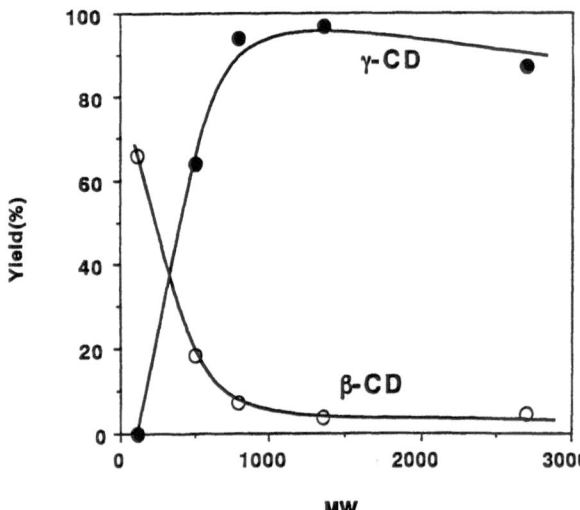

Figure 20.

of PIB with these CDs depends on the PIB molecular weight. Interestingly, the dependence is entirely different for the two CDs. The complexation with γ-CD does not occur for the monomer or the dimer and becomes almost quantitative at molecular weights exceeding 1000. On the other hand, the complexation with β-CD shows no such chain length selectivity. The ^{1}H NMR spectra as well as the yield data showed that the complex formation of PIB with γ-CD obeys the 3:1 stoichiometry, i.e., three isobutylene units are included in one γ-CD, in conformity with the geometry that the length of the former is nearly equal to the depth of the latter's cavity

The crystalline complex of PIB with either β or γ-CD was insoluble even in boiling water, but it was solubilized when urea was added to the suspension and heated. This fact demonstrates the important role played by hydrogen bonding played in stabilizing the complex. The X-ray diffraction pattern for the complex of PIB with γ-CD was entirely different from that for uncomplexed γ-CD and suggested an extended column structure.

Molecular models indicate that the PIB chain is able to penetrate the γ-CD cavity, but not the α-CD one, owing to the steric hindrance of the dimethyl groups on the main chain. This indication is again in agreement with the experimental observation that PIB forms a complex with γ-CD but not with α-CD. Another feature of the models is that a single γ-CD cavity can accommodate three isobutylene units, supporting what was mentioned above.

4 Rotaxanes and Catenanes

In recent years, with increasing recognition of the roles played by specific noncovalent interactions in biological systems and chemical processes, the science of noncovalent assemblies- often called supramolecular science- has aroused considerable interest [76]. The remaining part of this article reviews some important studies made on rotaxane and catenane, two classic types of supramolecular structure.

Rotaxanes are the compounds consisting of noncovalent entities called "rotor" and "axle" [77]. Figure 21 illustrates them schematically. Initially, attempts were made to prepare them by statistical methods, so that the yields were necessarily very low [78–80]. Recently, methods have been proposed for their more efficient synthesis, with renewed interest in their unique structure and properties. Section 4.1 summarizes some of the typical results obtained.

Catenanes are another example of noncovalent assemblies, consisting of two or more ring molecules interlocked together (see Fig. 21). They offer chemists an equally intriguing object of research. In Sect. 4.2, some recent attempts to synthesize them are reviewed.

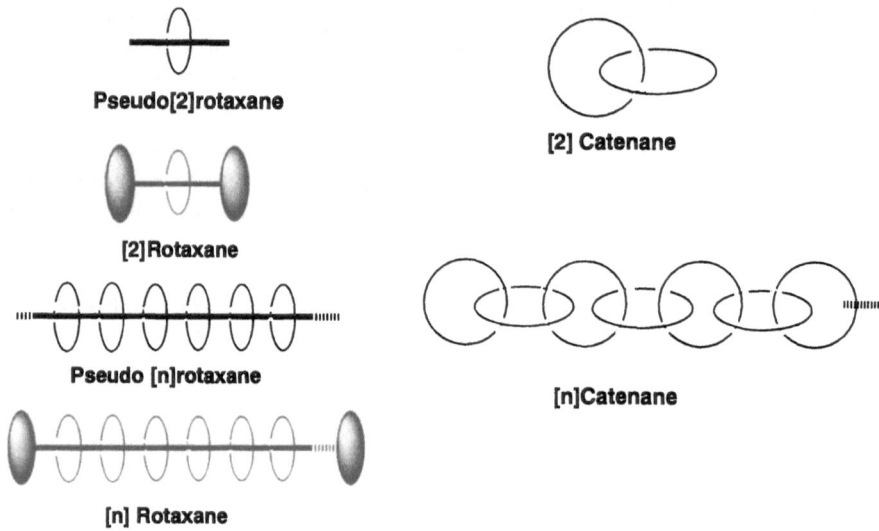

Figure 21.

4.1 Rotaxanes

4.1.1 Preparation of Rotaxanes

If in a rotaxane consisting of rings and a chain molecule the ends of the latter are not blocked by bulky substituents, the rings can slip off the chain. In this case, the rotaxane is called pseudo-rotaxane. The rotaxane containing a single ring is referred to as [2]rotaxane, and that containing $n + 1$ rings ($n \geqslant 1$) as [n]rotaxane. When n is arbitrarily large, the rotaxane is named polyrotaxane. The rings in rotaxanes are conventionally called "beads". We often use this naming in the following description.

In principle, a rotaxane can be obtained by end-capping an axle containing one or more rings with large substituents in ordered environments of non-covalent templating forces, in such a way that the order originally created by weak interactions is retained [81]. Thus, Allwood et al. [82] and Ashton et al. [83] prepared a pseudo-rotaxane shown in Fig. 22 by utilizing the fact that bis-p-phenylene-34-crown-10 forms a one-to-one complex with paraquat and 4,4'-bipyridinium dication derivatives in which the bipyridine nitrogen atoms are substituted by 2-hydroxyethyl and 2-(2-hydroxyethoxy) ethyl groups. Recently, Ashton et al. [84, 85] succeeded in obtaining a [2]rotaxane, shown in Fig. 23, by slippage reactions which allow a crown bead to change into a dumb-bell that consists of an axle molecule with large stoppers at its ends. The group of Stoddart and coworkers [86–88] found inclusion complex formation of cyclobis(paraquat-p-phenylene) with dimethoxybenzene and dimethoxynaph-thalene. Figure 24 shows the pseudo-rotaxane they obtained, in which

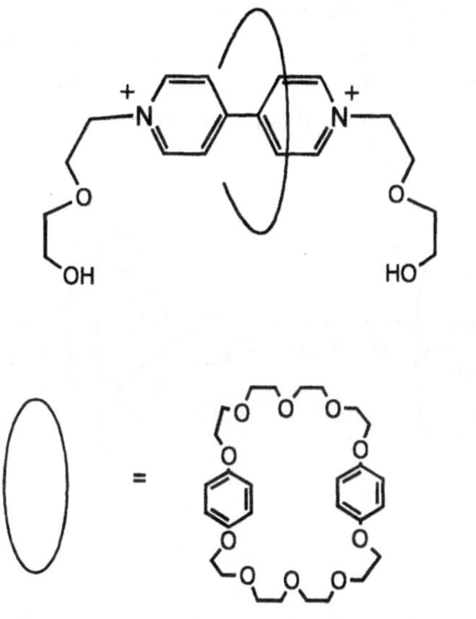

Figure 22.

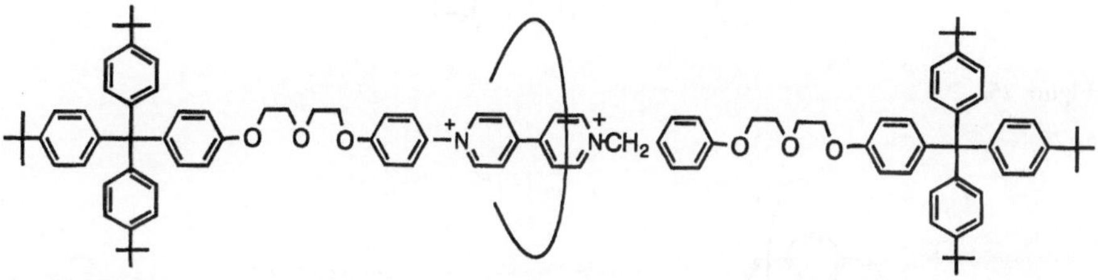

Figure 23.

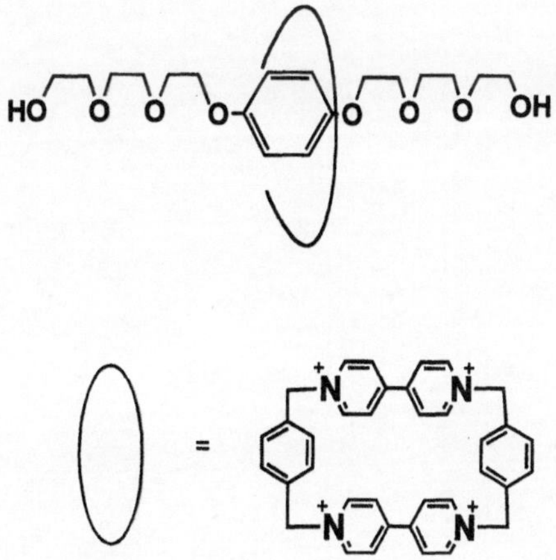

Figure 24.

cyclobis(paraquat-*p*-phenylene) is threaded by a hydroquinone derivative flanked by polyether chains [89]. The same group [81] substituted the chain ends with bulky groups such as triisopropyl silyl to obtain [2]rotaxanes. Ashton et al. [90, 91] prepared a pseudo-rotaxane containing two cyclobis(paraquat-*p*-phenylene)

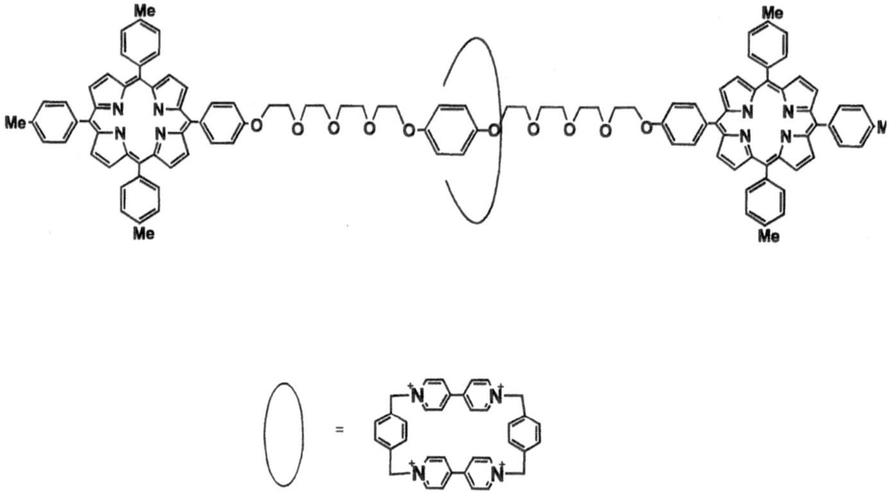

Figure 25.

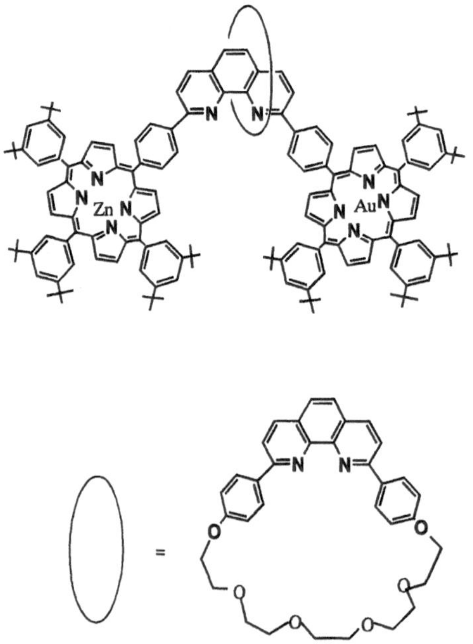

Figure 26.

beads threaded by a long polyether chain and converted it to a [2]rotaxane by using porphyrins as stoppers (Fig. 25). Finally, Fig. 26 shows the rotaxane obtained by Sauvage et al. [92], who used a copper(I)-based template technique. It contains two rigidly held porphyrins as stoppers.

4.1.2 Molecular Shuttles

Figure 27 illustrates a rotaxane of Stoddart et al. [93], in which the axle contains two "stations" interacting with the cyclobis(paraquat-p-phenylene) bead. Its ^{1}H NMR spectrum at room temperature indicated that the bead moves back and forth like a shuttle between the stations about 1800 times a second. Stoddart et al. [94, 95] prepared a [2]rotaxane containing two bipyridinium units and a crown, as shown in Fig. 28. The shuttling speed of the bead estimated by

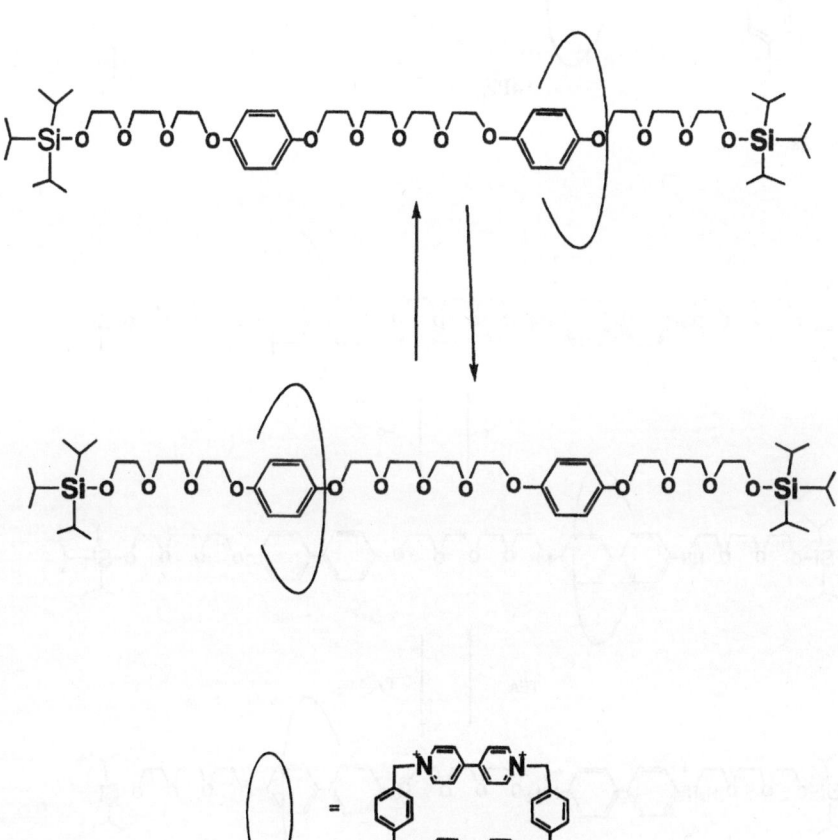

Figure 27.

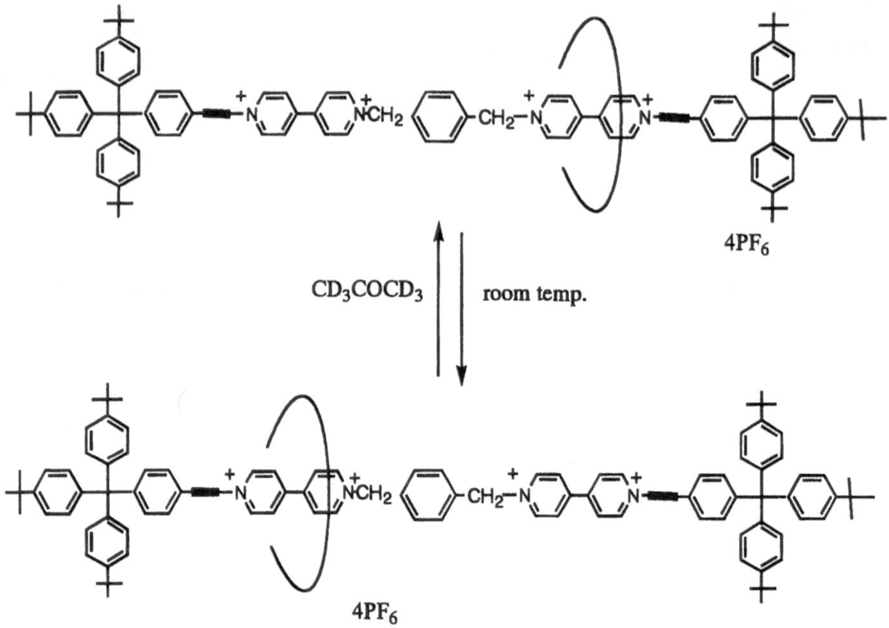

Figure 28.

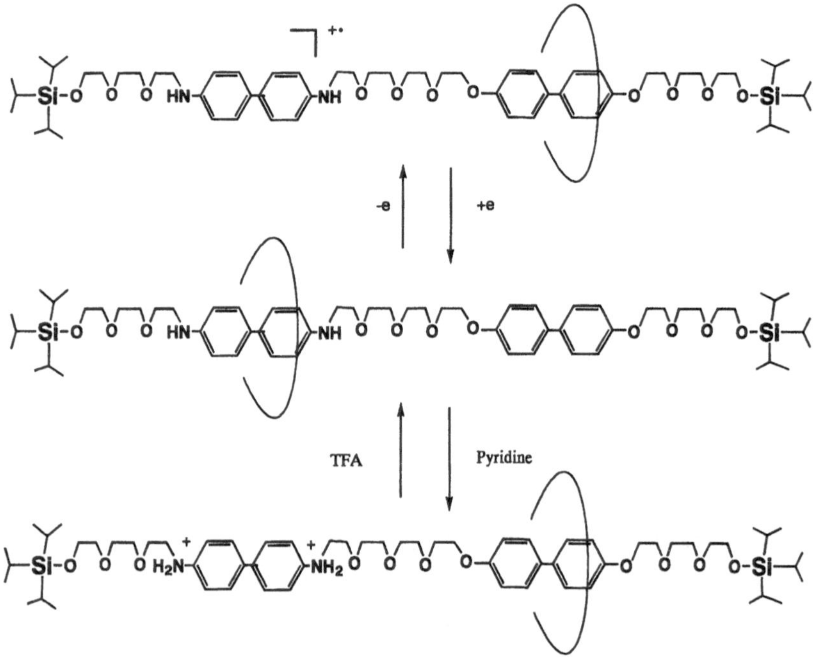

Figure 29.

^{1}H NMR was as fast as 300 000 times a second. It is interesting if the shuttling speed can be controlled by light or an electrochemical method [96]. Bissell et al. [97] obtained the molecular shuttle shown in Fig. 29, in which benzidine and bisphenol units act as the stations. At 229 K the tetracation bead was found to stay on the benzidine side at a probability of 86%, but when the compound was treated with an acid or oxidized electrochemically it turned out that the bead can move to the bisphenol side at a higher probability.

4.1.3 [2]Rotaxanes Containing Cyclodextrins

[2]Rotaxanes containing cyclodextrins and their derivatives as beads have been reported [98]. Some examples are as follows. Figure 30 shows the one prepared by Ogino [99] and by Ogino and Ohata [100], who used a cobalt complex for the stoppers. The [2]rotaxanes prepared by the group of Lawrence and coworkers contained biphenyl or porphyrin as the axle and dimethyl β-CD as the bead (Fig. 31) [101–103]. Figure 32 shows a pair of asymmetric [2]rotaxanes reported by Isnin and Kaifer [104], who used ferrocene and naphthalene sulfonate as stoppers. Wiley and Macartney [105] obtained symmetric [2]rotaxanes with pentacyanoiron complexes as stoppers and α-CD as a bead. Figure 33 illustrates one of them; more are shown in Figs. 34 and 35. The hydrophobic [2]rotaxane in Fig. 34, due to Wenz et al. [106], contains

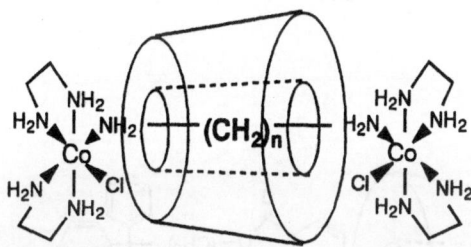

Figure 30.

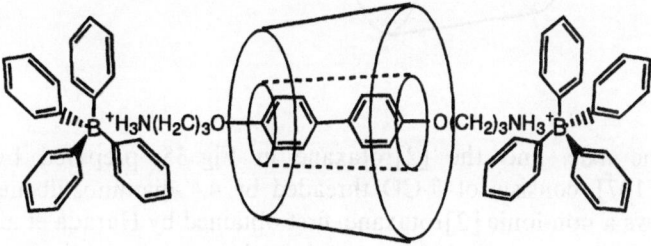

Figure 31.

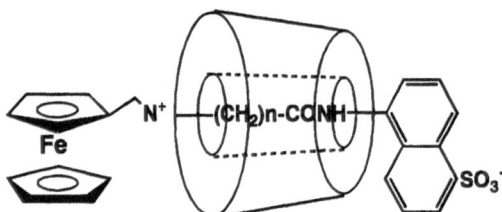

Figure 32.

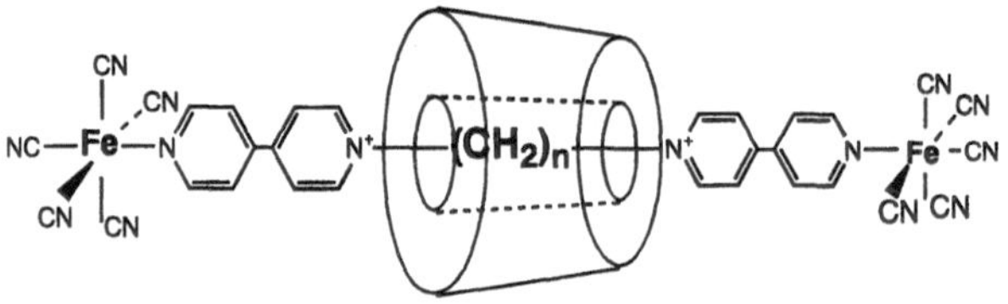

Figure 33.

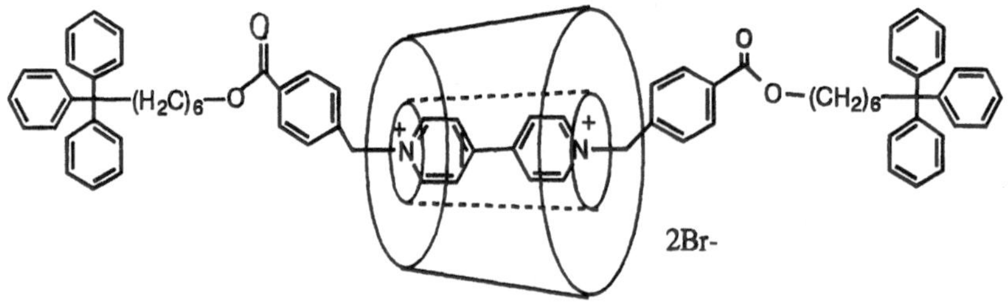

Figure 34.

bipyridinium in the axle, and the [2]rotaxane in Fig. 35, prepared by Nakashima et al. [107], consists of β-CD threaded by 4,4'-diaminostilbene. Finally, Fig. 36 shows a non-ionic [2]rotaxane, first obtained by Harada et al. [108], with a CD derivative as a bead and trinitrobenzene derivatives as stoppers.

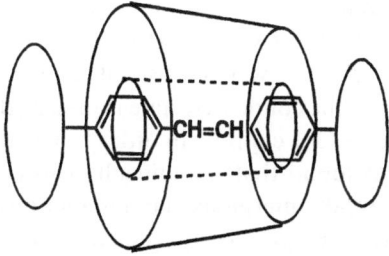

Figure 35.

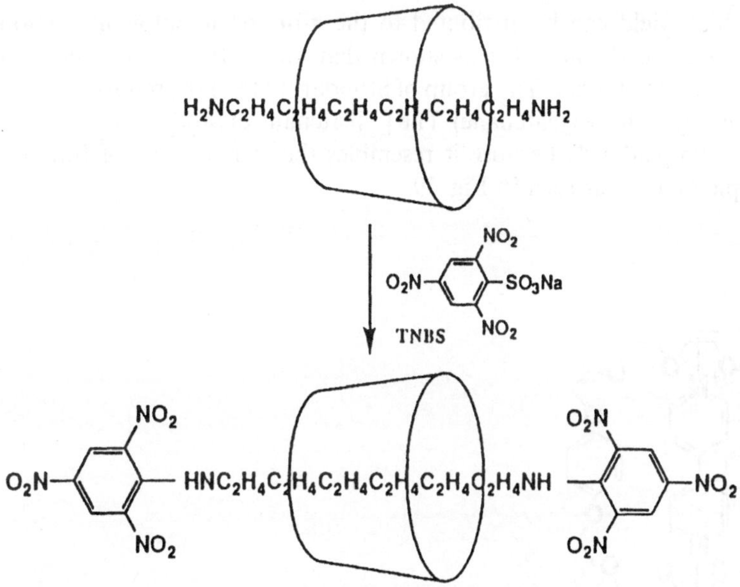

$$H_2NC_2H_4C_2H_4C_2H_4C_2H_4C_2H_4C_2H_4NH_2$$

$$\downarrow \quad \begin{array}{c} NO_2 \\ O_2N-\bigcirc-SO_3Na \\ NO_2 \end{array}$$

$$\text{TNBS}$$

$$O_2N-\begin{array}{c} NO_2 \\ \bigcirc \\ NO_2 \end{array}- HNC_2H_4C_2H_4C_2H_4C_2H_4C_2H_4C_2H_4NH -\begin{array}{c} O_2N \\ \bigcirc \\ O_2N \end{array}-NO_2$$

Figure 36.

4.1.4 Polyrotaxanes Consisting of Crown Ethers and Polymers

Gibson et al. [109] and Sjen et al. [110] reported pseudo-polyrotaxanes and polyrotaxanes consisting of crown ethers with various polymers. The resulting polyrotaxanes were nonstoichiometric. Their properties – including solubility and glass transition temperatures – were different from those of the starting polymers.

4.2 Catenanes

4.2.1 Preparation of Catenanes

As mentioned above, catenanes consist of two or more chemically nonbonded ring molecules interlocked with one another. A catenane consisting of n rings is referred to as [n]catenane.

Initial attempts to prepare catenanes aimed to thread a linear molecule (axle) through a ring molecule (ring) and "clip" the former to a ring. However, since this method is basically statistical, the yields of product were very small (actually less than 0.001%) [111]. It should be noted that the condensed state is preferred for threading, while dilute solutions are preferred for clipping in order to avoid oligomer/polymer formation. Recently, it was found that the difficulty can be overcome by taking advantage of specific weak intermolecular interactions between the precursors (i.e. the ring and the axle), and many catenanes have been successfully produced at high yields. Typical examples are as follows.

Stoddart and coworkers [112–115] cyclized bis(bipyridyl)s in the presence of crown ethers to obtain a [2]catenane at a yield of 70%. It is illustrated in Fig. 37. The high yield can be attributed to the π-donor-acceptor interaction between the ring and the axle. It was shown that one of the two intercrossed rings moves around the other. The group of Stoddart [116–118] reported [3–5] catenanes (see Fig. 38 for [3]catenane). The [5]catenane of Stoddart et al. [118] was named "olympiadane", because it resembles the symbol logo of International Olympic Games, as seen in Fig. 39.

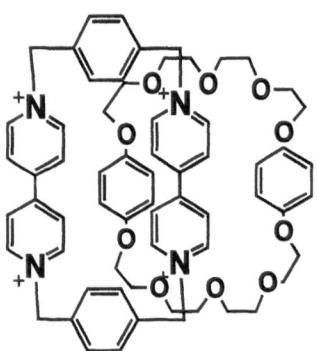

Figure 37.

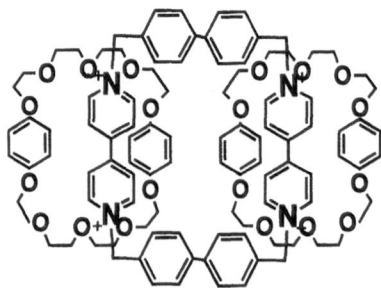

Figure 38.

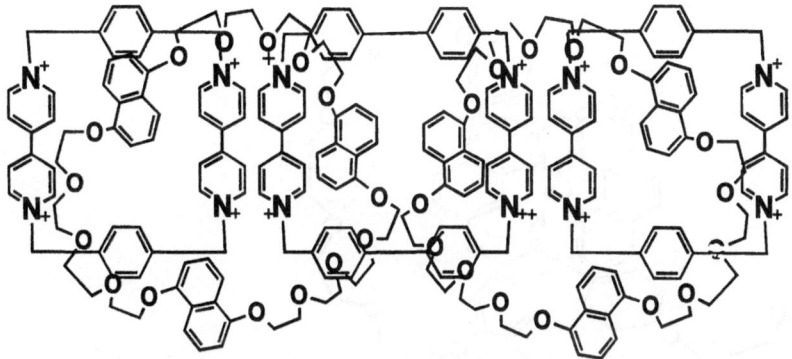

Figure 39.

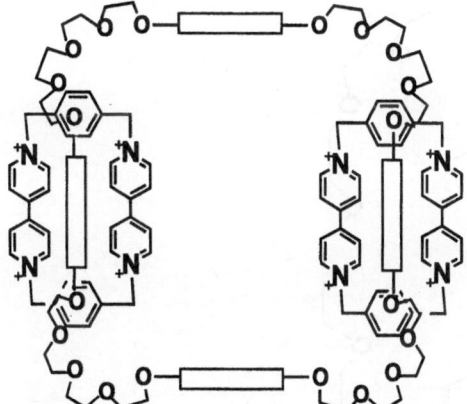

Figure 40.

4.2.2 Molecular Train

For a [2]catenane containing a smaller bipyridinium ring in a bigger ring made of four hydroquinones in acetonitrile at 80 °C it was shown [119] that the former moves around the latter 14×10^3 times a second and self-revolves 9×10^6 times a second. In the [3]catenane shown in Fig. 40, where one more bipyridinium ring is incorporated, the two smaller rings were found to move around without collision 10×10^3 times a second as though they were trains as a molecular scale [119].

4.2.3 Catenanes Containing Metal Complexes [120]

The group of Sauvage et al. [121] produced various [2]catenanes by using transition metal complexes as templates. Figure 41 illustrates one of their

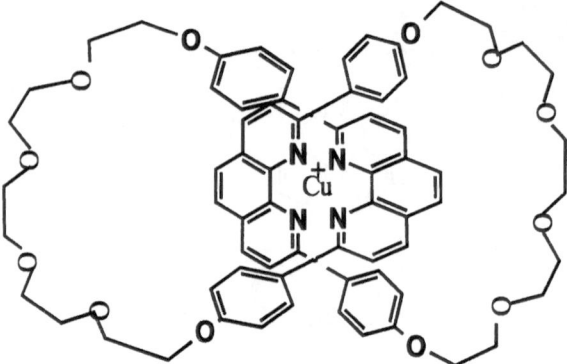

Figure 41.

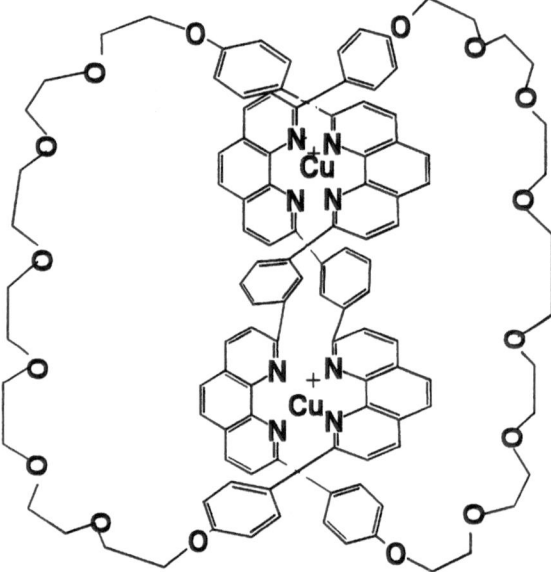

Figure 42.

products. The same group [122] obtained even a trefoil in which a ring is entangled itself (Fig. 42). The use of metal complexes was also made by Bitsch et al. [123] to obtain multi-ring catenanes, and Fujita et al. [124] produced a [2]catenane containing a cyclic Pd complex by a clipping method.

4.2.4 Catenanes Containing Cyclodextrin

Long ago Luttringshous et al. [125] tried to obtain a [2]catenane by clipping a dithiol thread penetrating a CD ring (see Fig. 43) by means of oxidation, but

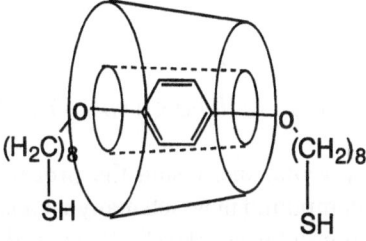

Figure 43.

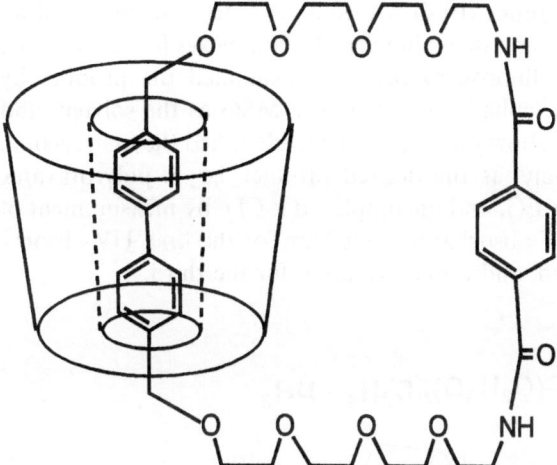

Figure 44.

they were unsuccessful. Recently, Stoddart et al. [126] successfully clipped a biphenyl derivative passing through a β-CD cavity to obtain the [2]catenane shown in Fig. 44. Harada et al. [127] obtained [2]catenanes by cyclizing a methylene chain penetrating β-CD with oligoethylene glycol derivatives. Figure 44 illustrates their product.

4.2.5 Other Catenanes

Vogtle et al. [128] obtained [2]catenanes containing cyclic lactams at high yields, and also one containing an azobenzene moiety, with the finding that the movement of the ring can be controlled by light [129]. Catenanes consisting of bipyridium cyclophanes and porphyrins have been reported by Gunter and Johnston [130].

5 Molecular Necklaces (Polyrotaxanes)

5.1 Pyrotaxanes Containing Polydisperse Poly(ethylene glycol) [131]

Harada et al. were the first to synthesize a polyrotaxane. Using the process shown in Scheme 1, they obtained an inclusion compound in which many α-CDs are threaded by a PEG chain and named it "molecular necklace". Wenz et al. [132] reported polyrotaxanes containing polyamines and α-CDs. Because of its significance and interest, the approach used by Harada et al. to obtain the molecular necklace is worth reproducing here in some detail.

Harada et al. started from preparing inclusion complexes by adding an aqueous solution of PEG bisamine (PEG-BA) to a saturated aqueous solution of α-CD at room temperature and then allowing the complexes formed to react with an excess of 2,4-dinitrofluorobenzene. They examined the product by column chromatography on Sephadex G-50, with DMSO as the solvent, and obtained the elution diagram shown in Fig. 46. They identified the first, second, and third fraction, respectively, as the desired product, i.e., a polyrotaxane, dinitrophenyl derivatives of PEG, and uncomplexed α-CD, by measurement of both optical rotation and UV absorbance at 360 nm for the first, UV absorbance at 360 nm for the second, and optical rotation for the third.

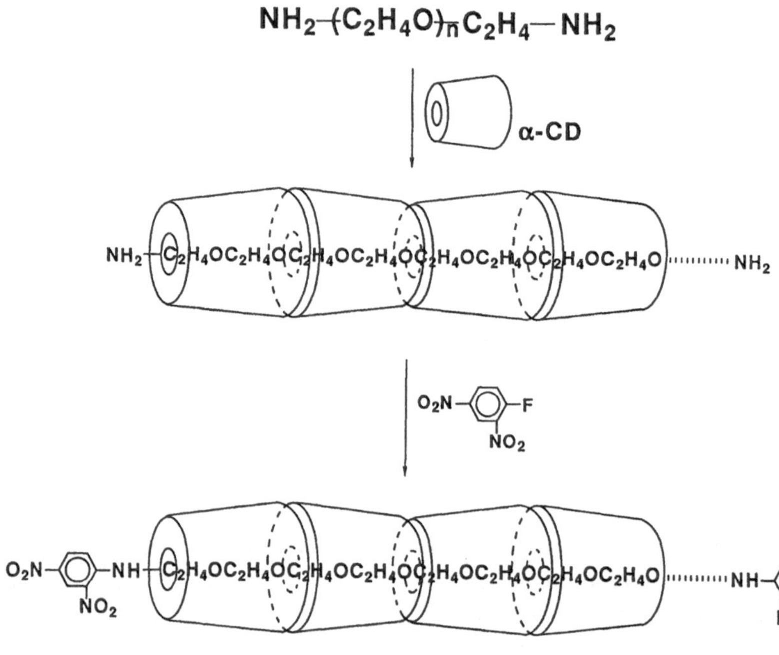
$NH_2(C_2H_4O)_nC_2H_4-NH_2$
α-CD
NH_2
$C_2H_4OC_2H_4OC_2H_4OC_2H_4OC_2H_4OC_2H_4OC_2H_4OC_2H_4O$
NH_2
O_2N
F
NO_2
O_2N
NH
NO_2
NH
NO_2
NO_2

Scheme 1.

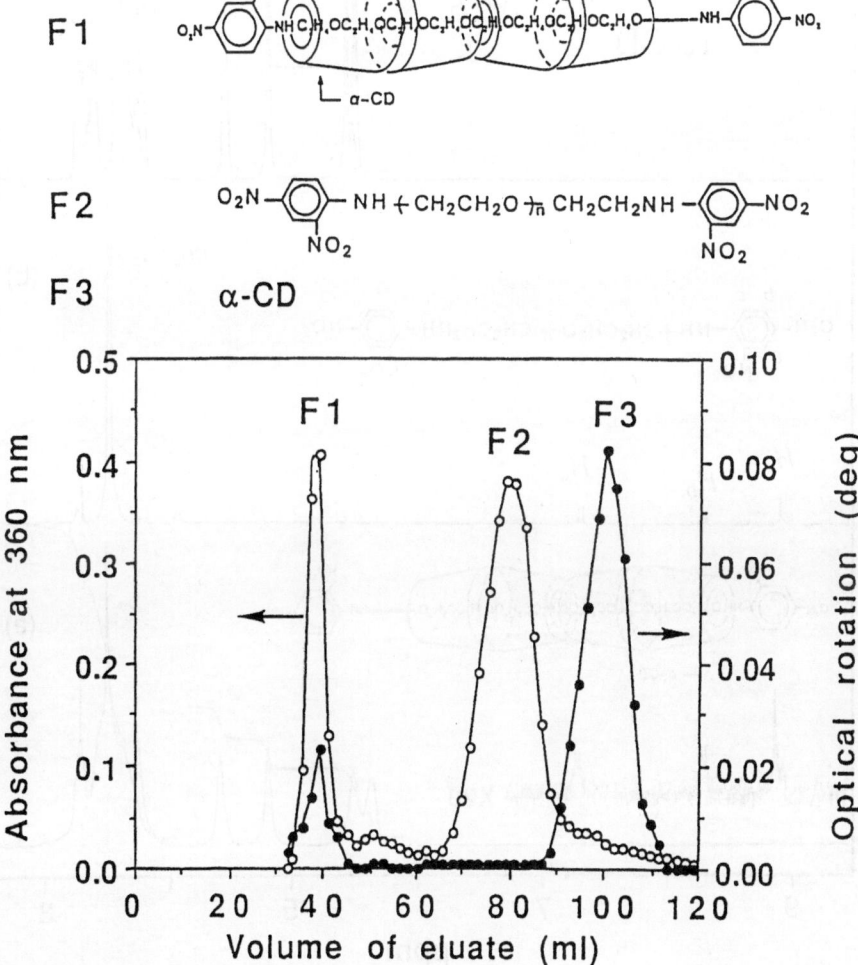

Figure 45.

Figure 46.

The polyrotaxane fraction was insoluble in water and DMF, but soluble in DMSO and 0.1 N NaOH. Its ^1H NMR spectrum- shown in Fig. 47, together with those for the second and third fraction- indicates that the polyrotaxane consists of CD, PEG-BA, and dinitrophenyl groups and that all peaks are broadened by complexation, which implies that α-CD rings become less mobile when including PEG. 2D NOESY NMR spectra showed that the H-3 and H-5

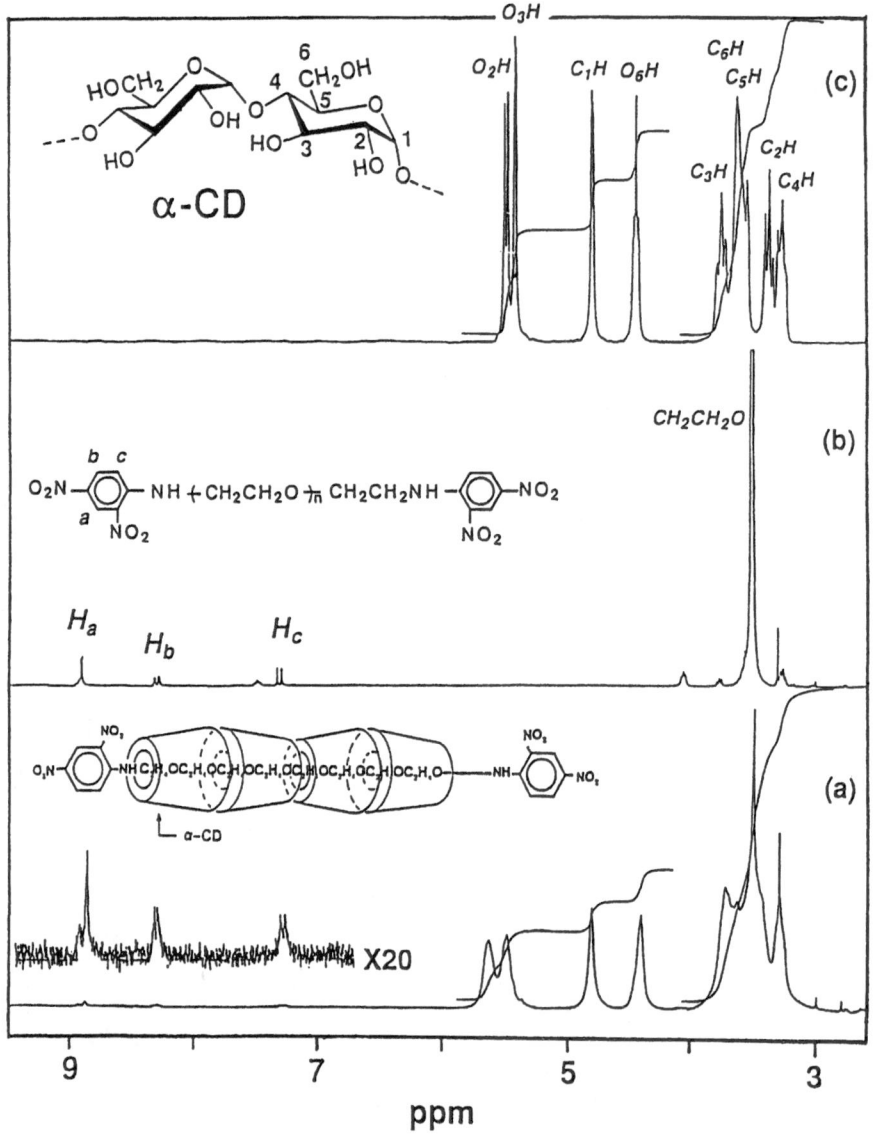

Figure 47.

Table 5. Polyrotaxanes prepared from PEG-BA with various molecular weights

Polyrotaxane	Molecular weight[b]	Number of ethylene glycol units (included + non-included)	Number of threaded α-CD[b]	Molar ratio of ethylene glycol units to α-CD
MN-1450	16500	33 (33 + 0	15	2.2
MN-2000	20000	45 (36 + 9)	18	2.5
MN*-2001[a]	19000	45 (34 + 11)	17	2.6
MN-3350	23500	77 (40 + 37)	20	3.9
MN-8500	44000	193 (72 + 121)	36	5.4
MN-20000	89000	454 (140 + 314)	70	6.5

[a] Prepared from Jeffamine ED-2001.
[b] Calcd. from UV-vis spectra, optical rotation, and ^{1}H NMR spectra.

protons of α-CD, which point to the CD cavity, can be correlated with the CH_2 of PEG, but the H-1, H-2, and H-4 protons of α-CD, which are located outside the CD cavity, do not. All these observations convinced Harada et al. that a polyrotaxane consisting of many α-CD rings threaded by a PEG chain was successfully obtained. The question is how many rings were trapped by a single chain.

Table 5 shows some characteristics of the polyrotaxanes prepared using PEG-BA of different molecular weights. As expected, the number of trapped CD rings increases as the PEG chain gets longer. The molar ratio of ethylene glycol units to α-CD is scattered around 2.4 for the three lowest molecular weights. This value is close to the stoichiometric ratio 2 – as was established earlier for the complex formation between PEG and α-CD –, so that, in these polyrotaxane samples, the rings cover essentially the entire length of the PEG chain. However, the molar ratios for the higher molecular weight samples are significantly larger than 2, indicating that the rings cover only part of the polymer chain, actually less than one half. Finally, the scanning tunnel micrographic image of a polyrotaxane, as presented in Fig. 49, makes it visible that several CDs are arranged along a polymer chain.

5.2 Polyrotaxane Containing Monodisperse Oligo (ethylene glycol) [133]

The PEG-BA samples used in the experiments mentioned above were polydisperse, so that the products were highly polydisperse and heterogeneous. To elucidate whether this fact was responsible for the change in the number of trapped CD rings with the PEG chain length, Harada et al. carried out an experiment using a monodisperse 28 mer PEG (MW = 1248), chosen because previous experience had shown that PEGs ranging between 1000 and 1500 in molecular weight most effectively complexed with α-CD.

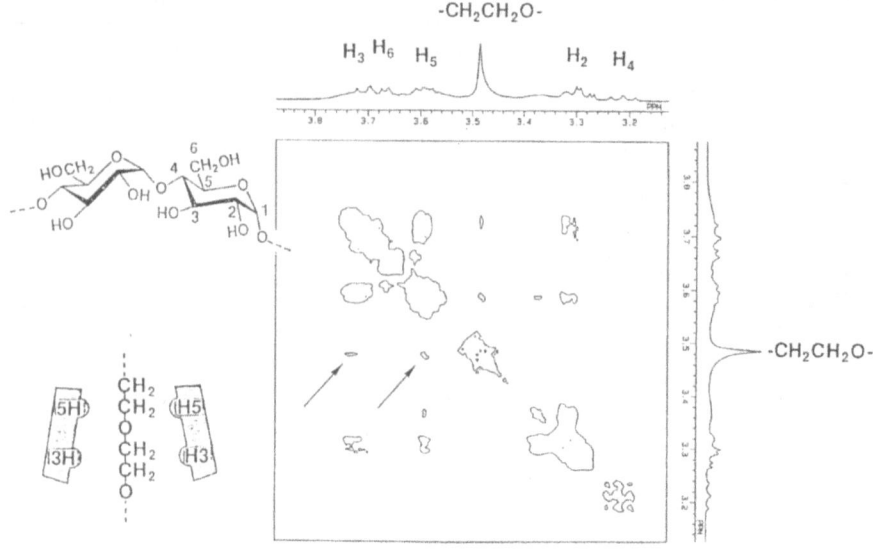

Figure 48.

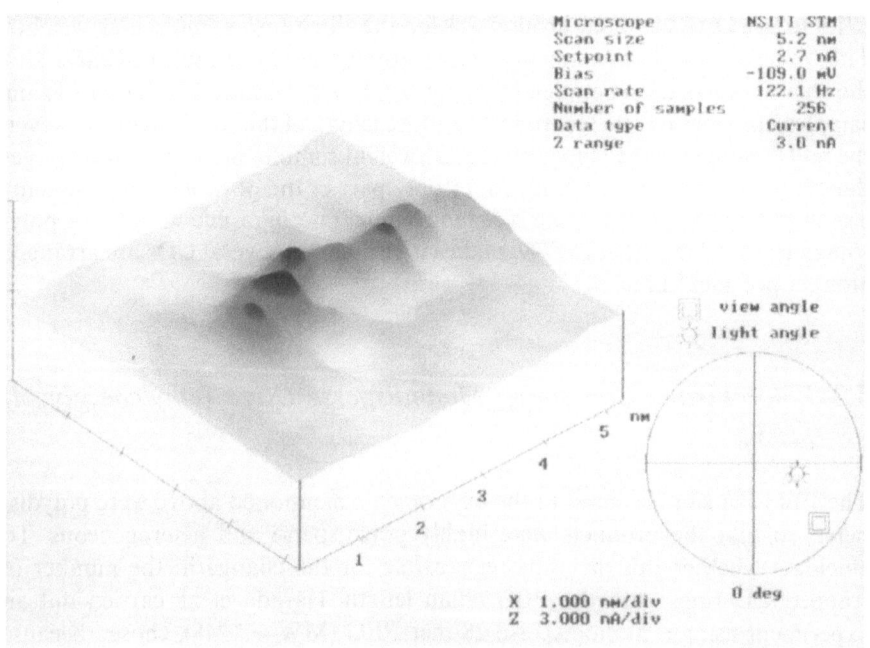

Figure 49.

The ^{13}C NMR spectrum of the polyrotaxane so obtained showed the peak for either C-4 or C-6 of CDs to appear as a doublet consisting of a broad peak in high magnetic field and a narrow peak in a lower field. Harada et al. assigned these broad and narrow peaks to the carbons of the CDs trapped by the polymer chain and those located at the chain ends, respectively, since the latter CDs should be more mobile than the former because they are free from hydrogen bonding, though they may receive some effect from the dinitrophenyl end groups.

These end groups were removed by cleaving the C–N bond with a strong base (25% NaOH) so that the CD rings may slip off the polymer chains, and the number of CDs per polyrotaxane was estimated by ^{1}H NMR, optical rotation, and UV absorption to be 12. This figure indicates that nearly the entire length of the 28 mer PEG chain was covered with CDs, because, as repeatedly noted above, the stoichiometric ratio for the complexation between PEG and α-CD is 2.

6 Molecular Tube [134–136]

Harada and coworkers proceeded further to obtain a tubular polymer from a PEG-α-CD polyrotaxane by using Scheme 2. The polyrotaxane was prepared

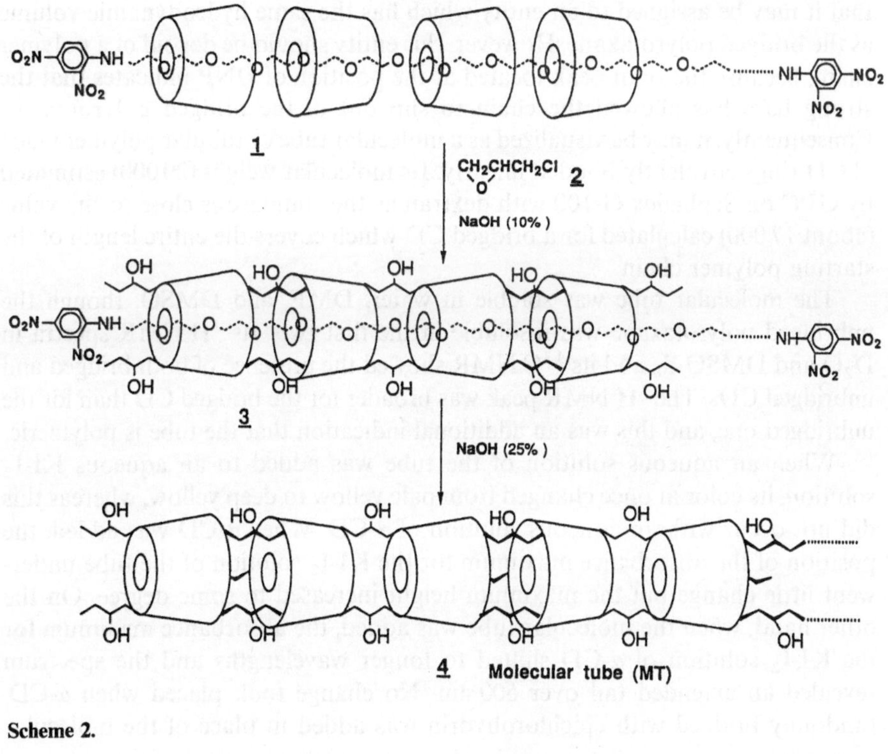

Scheme 2.

this time using a monodisperse PEG of 1450 molecular weight. From the above finding we may expect that when complexed with α-CD its length is almost entirely covered with CD rings.

First, the polyrotaxane was dissolved in 10% aqueous NaOH, and then epichlorohydrin was added to the suspension in order to bridge the CDs. After stirring for 36 hours at room temperature the reaction mixture was neutralized with HCl, and a yellow solid was precipitated from ethanol. The solid was dissolved in 25% aqueous NaOH with heating to remove the bulky dinitrophenyl(DNP) chain ends, and after cooling, the mixture was neutralized with HCl. Figure 50 shows the elution diagrams, measured by optical rotation (solid lines) and UV absorbance at 360 nm (dashed lines), for the following three mixtures: (a) a mixture of the unbridged polyrotaxane (molecular necklace) and 25% aqueous NaOH; (b) the neutral solution of the bridged polyrotaxane before treatment with 25% aqueous NaOH; (c) the reaction mixture of the bridged polyrotaxane with 25% aqueous NaOH.

In Fig. 50a, two peaks appear at the elution volumes expected for α-CD and DNP molecules, indicating that, with the strong base (25% NaOH), the DNP end groups were removed and the CD rings slipped off the polymer chain into the bulk solvent. Note that the resulting bare polymer chain is not observable by the methods used. The single peak in Fig. 50b can be identified with certainty as corresponding to the bridged CDs locking the polymer chain within them. The left peak in Fig. 50c is located at the same position as the peak in Fig. 50b, so that it may be assigned to an entity which has the same hydrodynamic volume as the bridged polyrotaxane. However, this entity should be devoid of a polymer chain, because the right peak located at the position of DNP indicates that the strong base has allowed the chain to slip out of the bridged polyrotaxane. Consequently, it may be visualized as a molecular tube or tubular polymer made of CD rings covalently bonded linearly. Its molecular weight (20 000) estimated by GPC on Sephadex G-100 with dextran as the standard is close to the value (about 17 000) calculated for a bridged CD which covers the entire length of the starting polymer chain.

The molecular tube was soluble in water, DMF, and DMSO, though the unbridged polyrotaxane was insoluble in the first two. Its ^{1}H NMR spectra in D_2O and DMSO-d_6 and its ^{13}C NMR showed the presence of both bridged and unbridged CDs. The ^{1}H NMR peak was broader for the bridged CD than for the unbridged one, and this was an additional indication that the tube is polymeric.

When an aqueous solution of the tube was added to an aqueous KI-I$_2$ solution, its color at once changed from pale yellow to deep yellow, whereas this did not occur with an aqueous solution of α-CD. When α-CD was added, the position of the absorbance maximum for the KI-I$_2$ solution of the tube underwent little change but the maximum height increased to some degree. On the other hand, when the molecular tube was added, the absorbance maximum for the KI-I$_2$ solution of α-CD shifted to longer wavelengths and the spectrum revealed an extended tail over 500 nm. No change took placed when α-CD, randomly bridged with epichlorohydrin was added in place of the molecular

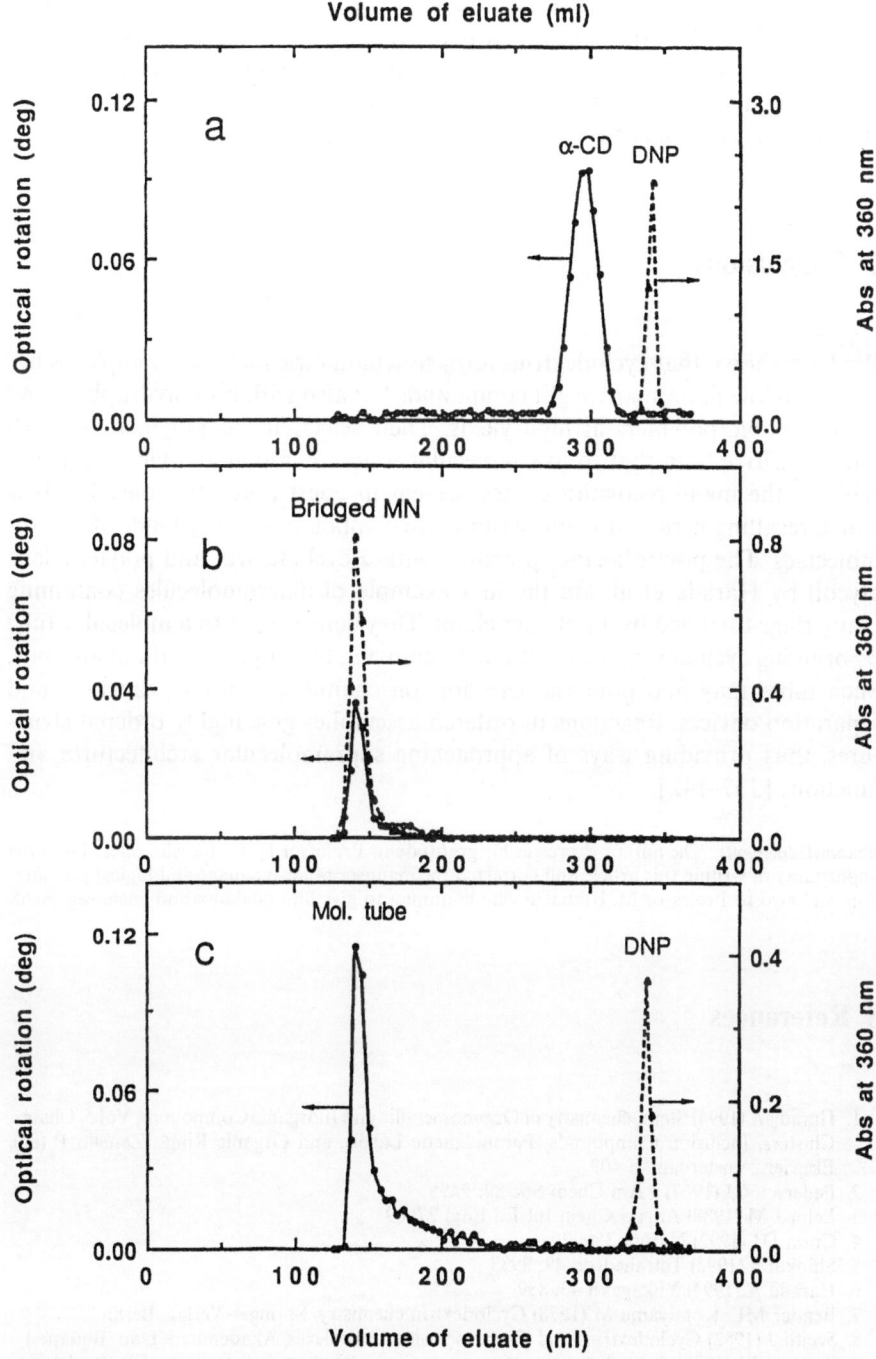

Figure 50.

tube. It follows from these observations that the I^{3-} ions are included linearly in the tube. The shift of the absorbance maximum was not as large as in the case of amylose and iodide, but similar to the case of amylopectin and polyiodide. Moreover, it was found that the spectrum underwent a maximum change when the molar ratio of CD to I^{3-} was $1:1$.

7 Conclusions

We have shown that cyclodextrins form stoichiometric inclusion complexes not only with low molecular weight compounds but also with both hydrophilic and hydro phobic polymers at high yields. Their selectivity to polymers is much more sensitive than that to low molecular weight compounds. This is attributable to the many recognition sites present in guest polymer molecules. It is worth recalling here that living systems are composed of many kinds of macromolecules. The polyrotaxanes prepared with α-cyclodextrin and poly(ethylene glycol) by Harada et al. are the first example of macromolecules containing many rings threaded by a polymer chain. They converted it to a molecular tube by bridging cyclodextrin rings and then removing the stoppers at the chain ends. Such tubes may find potential uses for ion channels, catalysts, capsules, and separation devices. Reactions in ordered assemblies give highly ordered structures, thus providing ways of approaching supramolecular architectures and functions [137–142].

Acknowledgements. The author expresses his gratitude to Professor H. Fujita who offered him the opportunity of writing this article and corrected the manuscript for language and logical presentation, and also to Professor M. Kamachi who continues to give him guidance and encouragement.

8 References

1. Harada A (1994) Stereochemistry of Organometallic and Inorganic Compounds, Vol 5, Chains, Clusters, Inclusion Compounds, Paramagnetic Labels, and Organic Rings. Zanello P (ed), Elsevier, Amsterdam, p 409
2. Pedersen CJ (1967) J Am Chem Soc 89: 2495
3. Lehn J-M (1988) Angew Chem Int Ed Engl 27: 89
4. Cram DJ (1992) Nature 356: 29
5. Shinkai S (1993) Tetrahedron 49: 8933
6. Harada A (1994) Yukagaku 43: 839
7. Bender ML, Komiyama M (1978) Cyclodextrin chemistry. Springer-Verlag, Berlin
8. Szejtli J (1982) Cyclodextrins and Their Inclusion Complexes, Akademiai Kiado, Budapest
9. Saenger W (1976) Jerusalem Symp Quantum Chem Biochem (ed) Pullman EB, Reidel Co., Dordrecht
10. Harada A, Takahashi S (1984) J Chem Soc, Chem Commun 645
11. Harada A, Takahashi S (1984) J Inclusion Phenom 2: 791

12. Harada A, Hu Y, Yamamoto S, Takahashi S (1988) J Chem Soc, Dalton Trans 729
13. Odagaki Y, Hirotsu K, Higuchi T, Harada A, Takahashi S (1990) J Chem Soc, Perkin Trans 1230
14. Shimada M, Harada A, Takahashi S (1991) J Chem Soc, Chem Commun 263
15. Harada A, Saeki K, Takahashi S (1987) J Inclusion Phenom 5: 601
16. Harada A, Saeki K, Takahashi S (1988) Carbohydr. Res 192: 1
17. Harada A, Takahashi S (1984) Chem Lett 2089
18. Harada A, Saeki K, Takahashi S (1985) Chem Lett 1157
19. Harada A, Takahashi S (1989) Organometallics 8: 730
20. Harada A, Takeuchi M, Takahashi S (1986) Chem Lett 1893
21. Harada A, Takeuchi M, Takahashi S (1986) Bull Chem Soc Jpn 61: 4367
22. Harada A, Takahashi S (1986) J Chem Soc, Chem Commun 1229
23. Harada A, Yamamoto S, Takahashi S (1989) Organometallics 8: 2560
24. Harada A, Takahashi S (1989) J Macromol Sci, Chem A26: 373
25. Harada A, Hu Y, Takahashi S (1986) Chem Lett 2083
26. Hu Y, Uno M, Harada A, Takahashi S (1990) Chem Lett 797
27. Hu Y, Uno M, Harada A, Takahashi S (1991) Bull Chem Soc Jpn 64: 1884
28. Hu Y, Harada A, Takahashi S (1990) J Mol Cat 60: L13
29. Hu Y, Harada A, Takahashi S (1990) Synt Commun 8: 1607
30. Harada A, Takahashi S (1988) J Chem Soc, Chem Commun 1352
31. Harada A, Shimada M, Takahashi S (1989) Chem Lett 275
32. Klingert B, Rihs (1990) Organometallics 9: 1135
33. Alston DR, Slawin AMZ, Stoddart JF, Williams DJ (1985) Angew Chem Int Ed Engl 24: 786
34. Harada A, Furue M, Nozakura S (1976) Macromolecules 9: 701
35. Harada A, Furue M, Nozakura S (1976) Macromolecules 9: 705
36. Harada A, Furue M, Nozakura S (1977) Macromolecules 10: 676
37. Harada A, Furue M, Nozakura S (1979) J Polym Sci. Polym Chem Ed 16: 189
38. Harada A, Furue M, Nozakura S (1980) Polym J 12: 29
39. Harada A, Furue M, Nozakura S (1980) Polym J 13: 777
40. Ogata N, Sanui K, Wada J (1979) J Polym Sci Polym Lett 14: 459
41. Maciejewski MM (1979) J Macromol Sci, Chem A13, 77: 1175
42. Kitano H, Okubo T (1977) J Chem Soc, Perkin II 432
43. Iijima T, Uemura T, Tsuzuku S, Komiyama J (1978) J Polym Sci, Polym Phys Ed 16: 793
44. Harada A (1993) Polym. News 18: 358
45. Harada A, Kamachi M (1990) Macromolecules 23: 2821
46. Harada A, Li J, Kamachi M (1993) Proc Jpn Acad 69: Ser. B 39
47. Harada A, Li J, Kamachi M (1993) Macromolecules 26: 5698
48. Harada A, Li J, Kamachi M (1992) Nature 356: 325
49. Harada A, Li J, Kamachi M (1993) Carbohydr Carbohydr Polym 25: 266
50. Harada A, Nakamitsu T, Li J, Kamachi M (1993) J Org Chem 58: 7524
51. Harada A (1996) Coord Chem Rev in press
52. Harada A (1995) Pharmacia 31: 1263
53. Harada A, Kamachi M (1994) J Chem Soc Jpn, Chem & Ind Chem 1994, 587
54. Harada A (1994) Seisan to Gijutu 46: 62
55. Harada A (1994) Yukagaku 43: 839
56. Harada A (1995) Biohistory 3: 25
57. Harada A, Kamachi M (1992) Hyomen/Surface 30: 110
58. Harada A (1994) Hyomen/Surface 32: 125
59. Harada A, Li J, Kamachi M (1994) Macromolecules 27: 4538
60. Bomer B, Heitz W, Kern W (1970) J Chromatogr 53: 51
61. Harada A, Li J, Kamachi M (1994) Nature 370: 126
62. Li J, Harada A, Kamachi M (1994) Polym J 26: 1019
63. Harada A, Kamachi M (1990) J Chem Soc Chem Commun 1322
64. Harada A, Okada M, Li J, Kamachi M (1995) Macromolecules 28: 8406
65. Harada A, Li J, Kamachi M (1993) Chem Lett 237
66. Harada A, Okada M, Kamachi M (1996) Acta Polymerica, in press
67. Harada A, Okada M, Kamachi M (1995) Proc Bilat Symp Polym Mater Sci (1995) 180
68. Harada A, Suzuki S, Nakamitsu T, Okada M, Kamachi M (1995) Kobunshi Ronbunshu, 52: 594

69. Steiner T, Koellner G, Saenger W (1992) Carbohydr Res (1992) 228: 321
70. Harada A, Kamachi M (1994) J Synth Org Chem Jpn 52: 831
71. Li J, Harada A, Kamachi M (1994) Bull Chem Soc Jpn 67: 2808
72. Yonemura H, Saito H, Matsushima S, Nakamura H, Matsuo T (1989) Tetrahedron Lett 30: 3143
73. Saito H, Yonemura H, Nakamura H, Matsuo T (1990) Chem Lett 535
74. Watanabe M, Nakamura H, Matsuo T (1992) Bull Chem Soc Jpn 65: 164
75. Harada A, Suzuki S, Li J, Kamachi M (1993) Macromolecules 26: 5267
76. Lehn J-M (1992) Angew Chem Int Ed Engl 29: 1304
77. Schill G (1971) Catenanes Rotaxanes and Knots, Academic Press, New York
78. Agam G, Graiver D, Zilkha A (1976) J Am Chem Soc 98: 5206
79. Harrison IT, Harrison S (1967) J Am Chem Soc 89: 5723
80. Harrison IT (1972) J Chem Soc Chem Commun 231
81. Annelli PL, Ashton PR, Ballardini R, Balazani V, Delgado M, Gandolfi MT, Goodnow TT, Kaifer AE, Philp D, Pietraszkiewicz M, Prodi L, Reddington MV, Slawin AMZ, Spencer N, Stoddart JF, Vicent C, Williams DJ (1992) J Am Chem Soc 114: 193
82. Allwood BL, Spencer N, Zavareh HS, Stoddart JF, Williams DJ (1987) J Chem Soc Chem Commun 1064
83. Ashton PR, Philp D, Reddington MV, Slawin AMZ, Spencer N, Stoddart JF, Williams DJ (1991) J Chem Soc Chem Commun 1680
84. Ashton PR, Belohradsky M, Philp D, Stoddart JF (1993) J Chem Soc Chem Commun 1269
85. Ashton PR, Belohradsky M, Philp D, Stoddart JF (1993) J Chem Soc Chem Commun 1274
86. Odel B, Reddington MV, Slawin AMZ, Spencer N, Stoddart JF, Williams DJ (1988) Angew Chem Int Ed Engl 27: 1547
87. Ashton PR, Odell B, Reddington MV, Slawin AMZ, Stoddart JF, Williams DJ (1988) Angew Chem Int Ed Engl 27: 1550
88. Reddingon MV, Slawin AMZ, Spencer N, Stoddart JF, Vicent C, Williams DJ (1991) J Chem Soc Chem Commun 630
89. Anelli PL, Ashton PR, Spencer N, Slawin AMZ, Stoddart JF, Williams DJ (1991) Angew Chem Int Ed Engl 30: 1036
90. Ashton PR, Philp D, Spencer N, Stoddart JF (1991) J Chem Soc Chem Commun 1677
91. Ashton PR, Johnston MR, Stoddart JF, Tolley MS, Wheeler JW (1992) J Chem Soc Chem Commun 1128
92. Chambron J-C, Heitz V, Sauvage J-P (1992) J Chem Soc Chem Commun 1131
93. Anelli L, Spencer N, Stoddart JF (1991) J Am Chem Soc 113: 5131
94. Ashton PR, Philp D, Spencer N, Stoddart JF (1992) J Chem Soc Chem Commun 1124
95. Ballardini R, Balzani, Gandolfi VMT, Prodi L, Venturi M, Philp D, Ricketts HG, Stoddart JF (1993) Angew Chem Int Ed Engl 32: 1301
96. Benniston AC, Harriman A (1993) Angew Chem Int Ed Engl 32: 1459
97. Bissell RA, Cordova E, Kaifer AE, Stoddart JF (1994) Nature 369: 133
98. Stoddart JF (1992) Angew Chem Int Ed Engl 31: 846
99. Ogino H (1981) J Am Chem Soc 103: 1303
100. Ogino H, Ohata K (1984) Inorg Chem 23: 3312
101. Manka JS, Lawrence DS (1990) J Am Chem Soc 112: 2440
102. Rao TVS, Lawrence DS (1990) J Am Chem Soc 112: 3614
103. Dick DL, Rao TVS, Sukumaran D, Lawrence DS (1992) J Am Chem Soc 114: 2664
104. Isnin R, Kaifer AE (1991) J Am Chem Soc 113: 8188
105. Wylie RS, Macartney DH (1992) J Am Chem Soc 114: 3136
106. Wenz G, Bey E, Schmidt L (1992) Angew Chem Int Ed 31: 783
107. Kunitake M, Kotoo K, Manabe O, Muramatsu T, Nakashima N (1993) Chem Lett 1033
108. Harada A, Li J, Kamachi M, Polymer Preprint, in press
109. Gibson HW, Marand H (1993) Adv Mater 5: 11
110. Sjen YX, Xie D, Gibson HW (1994) J Am Chem Soc 116: 537
111. Frisch HL, Wasserman E (1961) J Am Chem Soc 83: 3789
112. Ashton PR, Goodnow TT, Kaifer AE, Reddington MV, Slawin AMZ, Spencer N, Stoddart JF, Vicdent C, Williams DJ (1989) Angew Chem Int Ed Engl 28: 1396
113. Amabilino DB, Ashton PR, Tolley MS, Stoddart JF, Williams DJ (1993) Angew Chem Int Ed Engl 32: 1297

114. Ashton PR, Philp D, Spencer N, Stoddart JF, Williams DJ (1994) J Chem Soc Chem Commun 181
115. Ashton PR, Brown CL, Chrystal EJT, Goodnow TT, Kaifer AE, Parry KP, Philp D, Slawin AMZ, Spencer N, Stoddart JF, Williams DJ (1991) J Chem Soc Chem Commun 634
116. Ashton PR, Brown CL, Chrystal EJT, Goodnow TT, Kaifer AE, Parry KP, Slawin AMZ, Spencer N, Stoddart JF, Williams DJ (1991) Angew Chem Int Ed Engl 30: 1039
117. Amabilino DB, Ashton PR, Reder AS, Spencer N, Stoddart JF (1994) Angew Chem Int Ed Engl 33: 433
118. Amabilino DB, Ashton PR, Reder AS, Spencer N, Stoddart JF (1994) Angew Chem Int Ed Engl 33: 1286
119. Pietraszkiewicz, Spencer N, and Stoddart JF (1991) Angew Chem Int Ed Engl 30: 1042
120. Buchecker COD, Sauvage J-P (1987) Chem Rev 87: 795
121. Chambron JC, Heitz V, Sauvage J-P (1993) J Am Chem Soc 115: 12378
122. Buchecker COD, Guilhem J, Pascard C, Sauvage J-P (1990) Angew Chem Int Ed Engl 29: 1154
123. Bitsch F, Buchecker COD, Khemiss A-K, Sauvage J-P, Dorsselaer AV (1991) J Am Chem Soc 113: 4023
124. Fujita M, Ibukuro F, Hagihara H, Ogura K (1994) Nature 367: 720
125. Luttringhous, Cramer F, Prinzbach H, Henglein FM (1958) Liebigs An Chem 613: 185
126. Armspach D, Ashton PR, Moore CP, Spencer N, Stoddart JF, Wear TJ, Williams DJ (1993) Angew Chem Int Ed Engl 32: 854
127. Harada A, Li J, Kamachi M, unpublished.
128. Vogtle F, Meier S, Hoss R (1992) Angew Chem Int Ed Engl 31: 1619
129. Vogtle F, Muller WM, Muller U, Bauer M, Rissanen K (1993) Angew Chem Int Ed Engl 32: 1295
130. Gunter MJ, Johnston MR (1992) J Chem Soc Chem Commun 1163
131. Harada A, Kamachi M (1994) Ordering in Macromolecular Systems, Springer-Verlag, Berlin, 69
132. Wenz G, Keller B (1992) Angew Chem Int Ed Engl 31: 197
133. Harada A, Li J, Kamachi M (1993) J Am Chem Soc 116: 3192
134. Harada A, Li J, Kamachi M (1993) Nature 364: 516
135. Harada A, Li J, Kamachi M (1995) Macromolecular Engineering, Ed, Mishra MK, Nuyken O, Kobayashi S, Yagci Y, and Sar B, p 127, Plenum Press, New York
136. Harada A, Li J, Kamachi M (1995) J Macromol Sci, Macromol Rep A32: 813
137. Harada A (1995) KOBUNSHI/High Polymers Japan 44: 390
138. Harada A, Kamachi M (1992) KOBUNSHI/High Polymers Japan 41: 814
139. Harada A, Kamachi M (1992) Kagaku/Chemistry 47: 634
140. Harada A (1992) Kagaku/Chemistry 47: 631
141. Harada A, Kamachi M (1995) Chem & Chem Ind Jpn 48: 899
142. Harada A, Kamachi M (1996) Supramolecular Science, in press

Editor: Prof. H. Fujita
Received July 1996

Author Index Volumes 101-133

Author Index Vols. 1-100 see Vol. 100

Adolf, D. B. see Ediger, M. D.: Vol. 116, pp. 73-110.

Aharoni, S. M. and *Edwards, S. F.*: Rigid Polymer Networks. Vol. 118, pp. 1-231.

Améduri, B., Boutevin, B. and *Gramain, P.*: Synthesis of Block Copolymers by Radical Polymerization and Telomerization. Vol. 127, pp. 87-142.

Améduri, B. and *Boutevin, B.*: Synthesis and Properties of Fluorinated Telechelic Monodispersed Compounds. Vol. 102, pp. 133-170.

Amselem, S. see Domb, A. J.: Vol. 107, pp. 93-142.

Andrady, A. L.: Wavelenght Sensitivity in Polymer Photodegradation. Vol. 128, pp. 47-94.

Andreis, M. and *Koenig, J. L.*: Application of Nitrogen-15 NMR to Polymers. Vol. 124, pp. 191-238.

Angiolini, L. see Carlini, C.: Vol. 123, pp. 127-214.

Anseth, K. S., Newman, S. M. and *Bowman, C. N.*: Polymeric Dental Composites: Properties and Reaction Behavior of Multimethacrylate Dental Restorations. Vol. 122, pp. 177-218.

Armitage, B. A. see O'Brien, D. F.: Vol. 126, pp. 53-58.

Arndt, M. see Kaminski, W.: Vol. 127, pp. 143-187.

Arnold Jr., F. E. and *Arnold, F. E.*: Rigid-Rod Polymers and Molecular Composites. Vol. 117, pp. 257-296.

Arshady, R.: Polymer Synthesis via Activated Esters: A New Dimension of Creativity in Macromolecular Chemistry. Vol. 111, pp. 1-42.

Bahar, I., Erman, B. and *Monnerie, L.*: Effect of Molecular Structure on Local Chain Dynamics: Analytical Approaches and Computational Methods. Vol. 116, pp. 145-206.

Baltá-Calleja, F. J., González Arche, A., Ezquerra, T. A., Santa Cruz, C., Batallón, F., Frick, B. and *López Cabarcos, E.*: Structure and Properties of Ferroelectric Copolymers of Poly(vinylidene) Fluoride. Vol. 108, pp. 1-48.

Barshtein, G. R. and *Sabsai, O. Y.*: Compositions with Mineralorganic Fillers. Vol. 101, pp.1-28.

Batallán, F. see Baltá-Calleja, F. J.: Vol. 108, pp. 1-48.

Barton, J. see Hunkeler, D.: Vol. 112, pp. 115-134.

Bell, C. L. and *Peppas, N. A.*: Biomedical Membranes from Hydrogels and Interpolymer Complexes. Vol. 122, pp. 125-176.

Bennett, D. E. see O'Brien, D. F.: Vol. 126, pp. 53-84.

Berry, G.C.: Static and Dynamic Light Scattering on Moderately Concentraded Solutions: Isotropic Solutions of Flexible and Rodlike Chains and Nematic Solutions of Rodlike Chains. Vol. 114, pp. 233-290.

Bershtein, V. A. and *Ryzhov, V. A.*: Far Infrared Spectroscopy of Polymers. Vol. 114, pp. 43-122.

Bigg, D. M.: Thermal Conductivity of Heterophase Polymer Compositions. Vol. 119, pp. 1-30.

Subject Index

Springer
and the
environment

At Springer we firmly believe that an
international science publisher has a
special obligation to the environment,
and our corporate policies consistently
reflect this conviction.
We also expect our business partners –
paper mills, printers, packaging
manufacturers, etc. – to commit
themselves to using materials and
production processes that do not harm
the environment. The paper in this
book is made from low- or no-chlorine
pulp and is acid free, in conformance
with international standards for paper
permanency.